3rd International Conference on Theoretical and Experimental Studies in Nuclear Applications and Technology (TESNAT 2017)

EPJ Web of Conferences Volume 154 (2017)

Adana, Turkey
10 - 12 May 2017

Editors:

Abdullah Aydin
Ismail Hakki Sarpun
Eyyup Tel
Abdullah Kaplan
Bayram Demir

ISBN: 978-1-5108-4885-6

Printed from e-media with permission by:

Curran Associates, Inc.
57 Morehouse Lane
Red Hook, NY 12571

proceedings.com

Some format issues inherent in the e-media version may also appear in this print version.

Printed by Curran Associates, Inc. (2017)

For additional information, please contact EDP Sciences – Web of Conferences at the address below.

EDP Sciences – Web of Conferences
17, Avenue du Hoggar
Parc d'Activité de Courtabœuf
BP 112
F-91944 Les Ulis Cedex A
France

Phone: +33 (0) 1 69 18 75 75
Fax: +33 (0) 1 69 28 84 91

contact-edps@webofconferences.org

Additional copies of this publication are available from:

Curran Associates, Inc.
57 Morehouse Lane
Red Hook, NY 12571 USA
Phone: 845-758-0400
Fax: 845-758-2633
Email: curran@proceedings.com
Web: www.proceedings.com

TABLE OF CONTENTS

TESNAT 2017

3rd International Conference on Theoretical and Experimental Studies in Nuclear Applications and Technology

TESNAT conference series aim to discuss and compare all applicable methods as are being applied at present in nuclear physics. The problems faced in these fields at present are focused in the development of new methods and in the improving of existing techniques to achieve an understanding of existing experimental data and in predicting with high reliability new properties and processes. This conference proposes to bring together all these related communities with the goal of creating an enriching dialog across the disciplines. The 2017 conference had given an overview on the theoretical and experimental challenges in nuclear physics and applications. The main topics of the workshops are computational nuclear physics, medical physics, Monte Carlo applications in Nuclear Physics and other applications of nuclear physics.

TESNAT 2017 attracted about 200 participants and during the conference 7 invited and 98 plenary talks as well as 105 posters were presented. We would like to thank all the participants as well as the members of the International Scientific Committee, the Local Organizing Committee, the sponsors: TAEK, TUBITAK and especially the hosting Cukurova University for the extremely warm atmosphere of the workshop. We would also like to thank invited speakers Prof. Dr. Mohamed BELGAID, Prof. Dr. Ahmet BOZKURT, Prof. Dr. A. Güneş TANIR, Dr. Ahmed AZBOUCHE, Prof.Dr. Virgilio CORRECHER, Prof. Dr. Gülsen ÖNENGUT, and Prof.Dr. Jameel-Un NABI for their supports and leaderships.

We would also like to thank Dr. Muhittin ŞAHAN for their supports, Dr. Ahmet BÜLBÜL and Dr. Ercan YILDIZ for their helps in organization and the staff of "European Physical Journal Web of Conferences" for their help.

MİTHAT ÖZSAN AMFİSİ

Statement of Peer review

In submitting conference proceedings to EPJ Web of Conferences, I certify to the Publisher that I adhere to the **Policy on Publishing Integrity** of the journal in order to safeguard good scientific practice in publishing.

1. All articles have been subjected to peer review administered by the proceedings editors.
2. Reviews have been conducted by expert referees, who have been requested to provide unbiased and constructive comments aimed, whenever possible, at improving the work.
3. Proceedings editors have taken all reasonable steps to ensure the quality of the materials they publish and their decision to accept or reject a paper for publication has been based only on the merits of the work and the relevance to the journal.

Title, date and place of the conference

TESNAT 2017,
10-12 MAY 2017
ADANA, TURKEY

Proceedings editor(s):

ABDULLAH AYDIN
İSMAİL HAKKI SARPÜN
EYYUP TEL
ABDULLAH KAPLAN
BAYRAM DEMİR

Date and editor's signature

13 SEPTEMBER 2017

DOI: 10.1051/epjconf/201715401001

Particle induced nuclear reaction calculations of Boron target nuclei

Eyyup Tel[1,*], *Muhittin* Sahan[1], *Ismail Hakki* Sarpün[2], *Yusuf* Kavun[3], *Ali Armagan* Gök[1], *Meltem* Poyraz[1]

[1]Osmaniye Korkut Ata University, Department of Physics, Osmaniye, Turkey
[2]Afyon Kocatepe University, Physics Department, Afyonkarahisar, Turkey
[3]Manisa Celal Bayar University, Department of Physics, Manisa, Turkey

Abstract. Boron is usable element in many areas such as health, industry and energy. Especially, Boron neutron capture therapy (BNCT) is one of the medical applications. Boron target is irradiated with low energy thermal neutrons and at the end of reactions alpha particles occur. After this process recoiling lithium-7 nuclei is composed. In this study, charge particle induced nuclear reactions calculations of Boron target nuclei were investigated in the incident proton and alpha energy range of 5–50 MeV. The excitation functions for ^{10}B target nuclei reactions have been calculated by using PCROSS Programming code. The semi-empirical calculations for (p,α) reactions have been done by using cross section formula with new coefficient obtained by Tel et al. The calculated results were compared with the experimental data from the literature.

1 Introduction

Boron found wide spread in the rocks and water on the surface of the earth is an important element and is often underutilized, trace mineral naturally present in certain foods. Boron is used effectively in medical treatment. It is preferred because it gives effective successful results in the treatment of cancer. In order to be successful, a sufficient amount of ^{10}B must be selectively delivered to all tumor cells and enough thermal neutrons must be absorbed to cause lethal damage from the ^{10}B(n,α)^{7}Li capture reactions [1-5]. After the absorption of the thermal neutron, two product nuclei (He and Li) occurs. Their ranges in tissue are as short as the diameter of a cell nucleus and whose LET (Linear Energy Transfer) values are very high. Therefore, energy is released in the tumor cell which is killed with high probability because of the high LET owing to the short absorption ranges so that neighboring cells are not damaged [1]. In this study, particle induced nuclear reactions calculations of boron target nuclei have been investigated in the incident proton energy range of 1–50 MeV. The excitation functions for ^{10}B target nuclei reactions (p,α), (p,n), (p,p) have been calculated using the semi-empirical formula.

2 Method

Cline [6], Ribansky [7] and Griffin [8] uses the PCROSS code that is unified model based on the solution of the master equation in the form proposed. The empirical cross section formula is including proton inelastic cross section and Coulomb effects of reactions induced by proton can be expressed suggested in many studies [9-26] as follows:

$$\sigma_{(p,\alpha)} = C\,\sigma_{p-ne}\,\sigma_{Coul}\exp(as) \qquad (1)$$

Where $\sigma_{p\text{-}ne}$ and σ_{Coul} (in unit of mb) are the proton non-elastic and Coulomb effect cross sections, respectively [9]. The coefficients C and a have been determined from only empirical formulae. Z is the proton number of target nuclei.

3 Results and Discussions

In this study, Particle Induced Nuclear Reactions Calculations of Boron Target Nuclei have been investigated in the incident proton energy range of 1–50 MeV. The (p,α), (p,n) and (p,p) reactions have been calculated via the semi-empirical formula for the excitation functions of ^{10}B target nuclei.

Full exciton model calculations have been done with PCROSS code [27] that used the initial exciton number as n_o=1 of 1 neutron and 0 hole [10-23]. Theoretical calculations of ^{10}B(p,α) equilibrium and pre-equilibrium reactions are compatible with each other as seen in Fig. 1.

Equilibrium and pre-equilibrium calculations are tended to increase up to 20 MeV but then tend to decrease in harmony. As seen in Fig. 2, ^{10}B(p,n) reactions calculations shows unstable increases and

* Corresponding author: eyuptel@osmaniye.edu.tr

decreases after in the neighborhood of 43 MeV the pre-equilibrium.

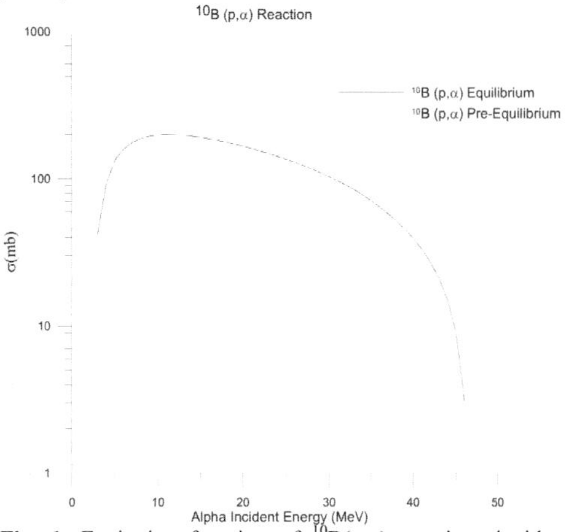

Fig. 1. Excitation function of $^{10}B(p,\alpha)$ reactions incident proton energy 1-50 MeV.

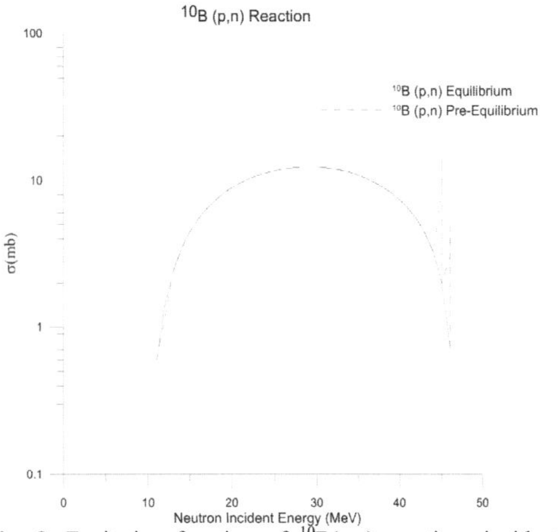

Fig. 2. Excitation function of $^{10}B(p,n)$ reactions incident proton energy 1-50 MeV.

In Fig. 3, ^{10}B (p,p) reactions calculations are starting to occur around 5 MeV and decreasing to until around 46 MeV.

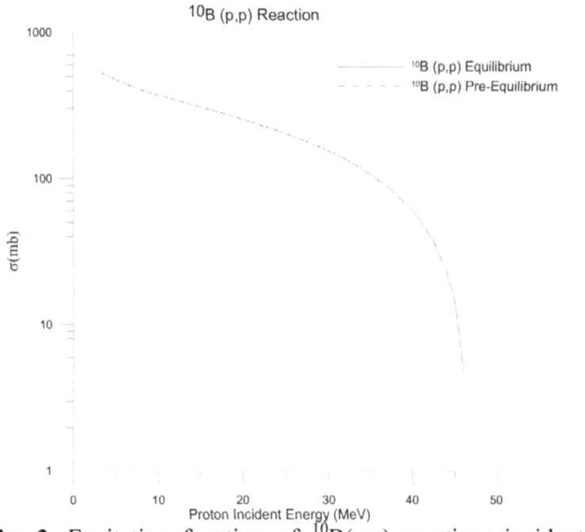

Fig. 3. Excitation function of $^{10}B(p,p)$ reactions incident proton energy 1-50 MeV.

Acknowledgment

The study was supported by Osmaniye Korkut Ata University (OKU) Science Research Projects Coordination Unit with the grant number: OKÜBAP-2017-PT3-022

References

1. R.F. Barth, J.A. Coderre, M.G. Vicente, T.E. Blue, Clin Cancer Res 2005, 11: 3987-4002.
2. V.Y. Butko, X. Chi, D.V. Lang, and A.P. Ramirez, Appl. Phys. Lett. https://doi.org/10.1063/1.163173683, 4773 (2003).
3. H. Klauk, M. Halik, U. Zschieschang, G. Schmid, W. Radlik, and W. Weber, J. Appl. Phys. https://doi.org/10.1063/1.151182692, 5259 (2002)
4. H. Tada, H. Touda, M. Takada, and K. Matsushige, Appl. Phys. Lett. https://doi.org/10.1063/1.12561476, 873 (2000)
5. R.J. Harris, W.B. Shuler, M. Eckhause, R.T. Siegel, R.E. Welsh, Phys. Rev. Lett. **20**, 505 (1968)
6. C.K. Cline, M. Blann: Nucl. Phys. A **172** 225 (1971)
7. I. Ribansky et al.: Nucl. Phys. A **205**545 (1973)
8. J.J. Griffin, Phys.Rev. Lett. **17** 478 (1966)
9. E. Tel, A. Aydın, E.G. Aydın, A. Kaplan, O. Yavas, I.A. Reyhancan, PRAMANA Indian Academy of Sciences **74**, 6, 931 (2010)
10. M. Sahan, E. Tel, H. Sahan, A. Kara, A. Aydin, A. Kaplan, I.H. Sarpun, B. Demir, S. Akca, E. Yildiz, J Fusion Energ **34,** 493 (2015).
11. J.M. Akkermans et al.: Phys. Rev. C **22** 73. (1980)
12. M. Belgaid, M. Asghar, Appl. Radiat. Isot. **49** 1497 (1998)
13. E. Tel, H.M. Sahin, A. Kaplan, A. Aydin, T. Altinok, Ann. Nucl. Energy **35** (2) 220 (2007)
14. M. Yiğit, E. Tel and I.H. Sarpün, Nuclear Instruments and Methods in Physics Research A **385**, 59 (2016)
15. E. Tel, M. Yiğit, G. Tanır, J. Fusion Energ. **31,** 184 (2012)
16. E. Tel, M. Yiğit, G. Tanır, J. Fusion Energ. **32,** 273 (2013)
17. M. Yiğit and E. Tel, Annals of Nuclear Energy **69,** 44 (2014)
18. E. Tel, S. Akca, A. Kara, M. Yiğit and A. Aydın, J. Fusion Energ. **32**, 531 (2013)
19. H. Şahan, E. Tel, M. Yiğit, J. Fusion Energ. **34**, 16 (2015)
20. M. Yiğit and E. Tel. Nuclear Engineering and Design **293**, 97 (2015)
21. M. Yiğit and E. Tel, J. Fusion Energ. **32**, 442 (2013)
22. M. Yiğit, J. Fusion Energ. **34**, 140 (2015)
23. M. Yiğit and E. Tel, Journal of Radioanalytical and Nuclear Chemistry **306,** 203 (2015).
24. A. Kaplan, J. Fusion Energ. **32**, 382 (2013).
25. A. Aydin, I.H. Sarpun, A. Kaplan, Phys. Atom Nucl. **77**, 321 (2014).
26. A. Aydin, H. Pekdogan, A. Kaplan, I.H. Sarpun, E. Tel, B. Demir, J. Fusion Energ. **34**, 1105 (2015).
27. R. Capote et al., Final Report on Research Contract 5472/RB, INDC(CUB)-004 (Higher Institute of Nuclear Science and Technology, Cuba). Translated by the IAEA on March 1991 (PCROSS program code) (1991).

Calculation of (n,α) reaction cross sections by using some Skyrme force parameters for Potassium (^{41}K) target nuclei

Eyyup Tel[1], *Muhittin* Sahan[1], *Hasancan* Alkanli[1], *Halide* Sahan[1], *Mustafa* Yigit[2]

[1]Osmaniye Korkut Ata University, Department of Physics, Osmaniye, Turkey
[2]Aksaray University, Department of Physics, Aksaray, Turkey

Abstract. In this study, the (n,α) nuclear reaction cross section was calculated for ^{41}K target nuclei for neutron and proton density parameters using SKa, SKb, SLy5, and SLy6 Skyrme force. Theoretical cross section for the (n,α) nuclear reaction was obtained using a formula constituted by Tel et al. (2008). Results are compared with experimental data from EXFOR. The calculated results from formula was found in a close agreement with experimental data.

1 Introduction

Potassium (with a symbol K, Z=19) presents in some fruits and vegetables such as peach and melons etc. [1-3]. In recent years, Potassium has been used in different area like medicine and agriculture etc. [1-5]. Nowadays, nuclear reactions have been using different cross section formula similar to Tel et al. formula [6]. The Hartree-Fock-Skyrme-Method is used for studying the properties and structure of nuclei [7-12]. In addition, many properties of nuclei are calculated using this method such as proton (ρ_p) and neutron (ρ_n) densities. In this calculation, we investigated the proton (ρ_p) and neutron (ρ_n) densities for ^{41}K target nuclei using the Skyrme-Hartree-Fock [8, 10, 11] calculation method with the SKa, SKb, SLy5, and SLy6 force parameters [7, 13, 14]. From these calculations, the new proton and neutron densities were obtained. The theoretical results calculated for proton (ρ_p) and neutron (ρ_n) densities were used in the formula given by reference [6] for the (n,α) nuclear reaction cross section at incident neutron energy of 14-15 MeV [6].

2 Results and Discussion

In this study, we calculated the (n,α) theoretical nuclear reaction cross section for ^{41}K target nuclei. We used SKa, SKb, SLy5, and SLy6, Skyrme force parameters for calculations [7, 13, 14]. SKa, SKb, SLy5, and SLy6, Skyrme force parameters were given in Table 1 and Fig. 1-4. These parameters were then used in the Skyrme-Hartree-Fock Program (HAFOMN) [11, 15]. Cross section calculations were obtained for target nucleus with radius of 1.8 fm and then were compared with the semi-empirical results constituted by Tel et al. formula [6]. For the mass numbers between 37 and 239, this formula is given as follows [6];

$$\sigma_{(n,\alpha)} = 16.15(A^{1/3}+1)^2 e^{-33.01s} \quad (1)$$

where A is mass number of atom, s is asymmetry parameter (S=(N-Z)/A).

Table 1. SKa, SKb, SLy5, and SLy6 Skyrme Force Parameters [7, 13, 14].

	SKa	SKb	SLy5	SLy6
t_0	-1602.78	-1602.78	-2484.88	-2479.50
t_1	570.88	570.88	483.13	462.18
t_2	-67.70	-67.70	-549.40	-448.61
t_3	8000	8000	13763	13673
t_4	125	125	126	122.00
x_0	-0.02	-0.02	0.778	0.825
x_1	0	-0.165	-0.328	-0.465
x_2	0	0	-1.00	-1
x_3	-0.286	-0.286	1.267	1.355
α	1/3	1/3	1/6	1/6

In earlier works, neutron and proton data for asymmetry parameter (S=(N-Z)/A) were used. But, for this study, we used in the formula developed by Tel et al. [6] proton and neutron density data for asymmetry parameter (S=(ρ_n-ρ_p)/(ρ_n+ρ_p)) [6,7,9]. Theoretical cross section values obtained with the cross section values ^{41}K(n,α)^{38}Cl results that are obtained using SKa, SKb, Sly5, and SLy6 parameters are given in Table 2 [7,13-14]. The neutron-number (N = 22) and neutron densities (ρ_n) are higher than proton-number (Z=19) and proton densities (ρ_p) because of Z=19 and N=22 for ^{41}K. The obtained value of the proton density (ρ_p) for ^{41}K target nuclei at the r = 1.8 fm have approximately been from 0.075 (for SKa and SLy5), 0.074 (for SKb), 0.079 (for SLy6) [7, 13, 14]. Moreover, value of the neutron density (ρ_n) for ^{41}K target nuclei at the r = 1.8 fm have approximately been from 0.081 (for SKa and SKb), 0.082 (for Sly5), 0.087 (for SLy6) and also obtained value of the asymmetry parameter for ^{41}K target nuclei at the 0.040 (for Ska), 0.043 (for SKb and SLy5), 0.044 (for Sly6) [7, 13, 14]. (see Figs. 1-4 and Table 2). Many experimental data were found from 1953 to 2017 for Potassium (for $^{41}K(n,\alpha)^{38}Cl$) [16]. Some experimental cross sections data were given in this study. For example; Garuska et al. found the cross section to be 30 ± 3 mb at 14.6 MeV neutron induced reactions [16, 17]. Filatenkov et al. found the cross section as to be 34.7 mb ± 1.6 mb at 14.1 ± 0.1 MeV neutron induced reactions [16-18]. Ercan et al. determined the experimental cross section as to be 36 ± 3 mb at 14.6 ± 0.1 MeV neutron induced reactions [16, 19]. Ikeda et al. found the cross sections to be 37.6 ± 2.8 mb at 13.97 MeV neutron induced reactions [16, 20]. Anders et al. found the cross section as to be 33 ± 1.3 mb at 14.7 ± 0.3 MeV neutron induced reactions [16, 21]. Bormann et al. found the cross section to be 12 ± 5 mb at 14.1 ± 0.05 MeV neutron induced reactions [16, 22]. Janczyszyn et al. found the cross section as to be 11 ± 2 mb at 14.0 MeV neutron induced reactions [16, 23].

Table 2 Theoretical cross section results for $^{41}K(n,\alpha)$ nuclear reactions for r=1.8 fm

SHF Parameters	Proton densities	Neutron densities	Asymmetry parameter	σ_{Theo} (mb)
SKa	0.075	0.081	0.040	10.08
SKb	0.074	0.081	0.043	9.086
SLy5	0.075	0.082	0.043	9.142
SLy6	0.079	0.087	0.044	9.046

We compared our data of target ^{41}K with literature data from EXFOR around 14-15 MeV [16, 18]. In this study, the obtained neutron and proton density results were depicted in Figs. 1-4. For neutron incident energy at 14.00 MeV, the experimental data is 11 ± 2 mb [23] and theoretically calculations are about 10.08 mb for SKa, 9.086 mb for SKb, 9.142 mb for SLy5, and 9.046 mb for SLy6 at r = 1.8 fm. These parameters were then used in the Skyrme-Hartree-Fock-program (HAFOMN) [7, 13-15]. Empirical results are found in compatible with theoretical data obtained in reference [6].

3 Conclusion

Many researchers have studied experimental and theoretical cross sections in recent years. In this study, (n,α) nuclear theoretical cross section reactions have been investigated for ^{41}K target nuclei incident neutron energy of 14-15 MeV. The attained data have also been contrasted on the existing some experimental values in EXFOR [16]. The attained theoretical and experimental results can be explained as follows;

In order to be calculate (n,α) reaction cross section, we used the formula developed by Tel et al. formula [6]. In Equation 1 developed by Tel et al. [6] can be used to calculate cross section with SKa, SKb, SLy5 and SLy6 Skyrme-force-parameters for ^{41}K target nuclei [7, 13-14]. The obtained results were compared with experimental result for 1.8 fm radius (see Figs. 1-4). In order to be calculate (n,α) different radius reaction cross section for ^{41}K target nuclei, we used Tel et al formula [6] and the we obtained theoretical cross section agreement with experimental results (see Figs. 1-4).

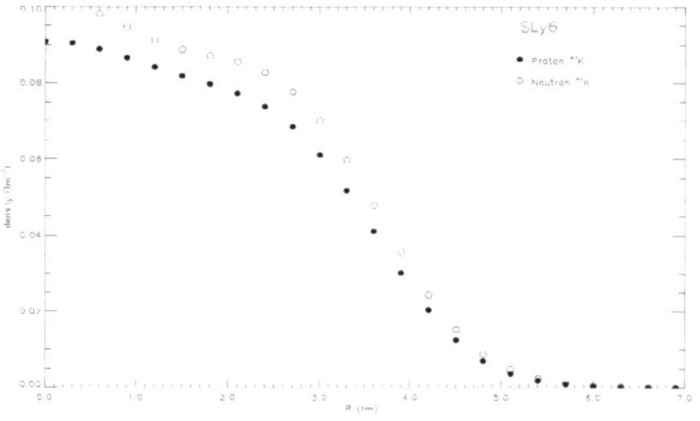

Fig. 1 $^{41}K(n,\alpha)^{38}Cl$ SLy6 proton and neutron density values.

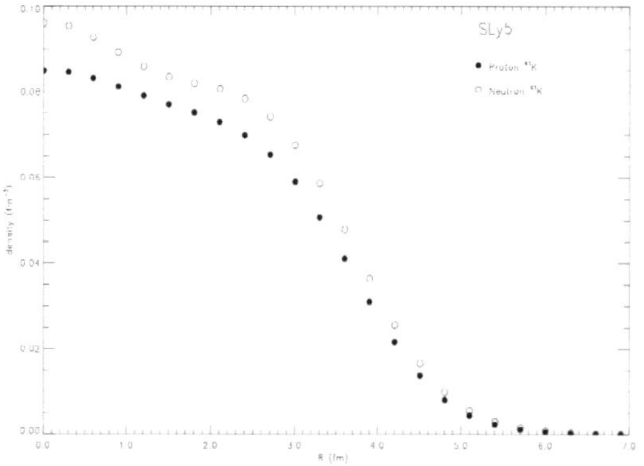

Fig. 2. $^{41}K(n,\alpha)^{38}Cl$ SLy5 proton and neutron density values

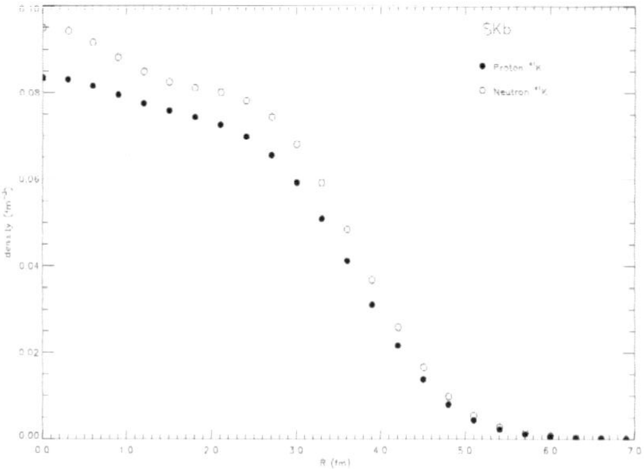

Fig. 3. $^{41}K(n,\alpha)^{38}Cl$ SKb proton and neutron density values

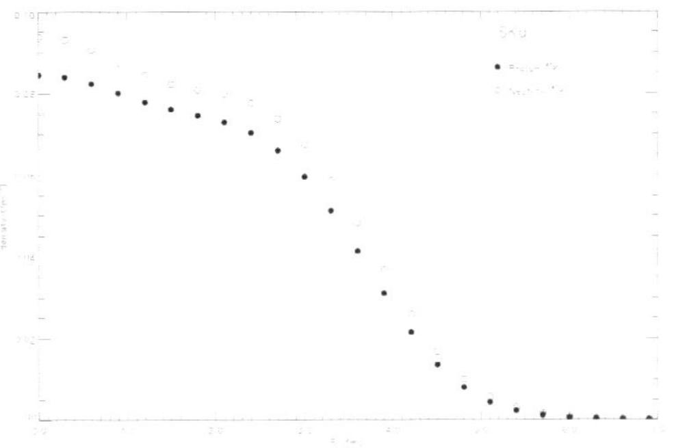

Fig. 4. $^{41}K(n,\alpha)^{38}Cl$ SKa proton and neutron density values

Acknowledgment

The study was supported by Osmaniye Korkut Ata University (OKU) Science Research Projects Coordination Unit with the grant number: OKÜBAP-2015-PT3-012.

References

1. J. S. Jones, J. Urol., **171 (1)**, 353 (2004)
2. S. Lamid, G. Klingbeil, Am. J. Clin. Nutr. **31(12)**, 2135 (1978)
3. S.K. Corbacioglu, S. Guler, D. Yagmur, V. Ulker, I. Kilicaslan, Turkish Journal of Emergency Medicine, **12(1)**, 41 (2012).
4. I. Sönmez, M. Kaplan, S. Sönmez, Journal of Batı Akdeniz Agricultural Research Inst.,**25(2)**, 24 (2008)
5. F. J. Gennari, New England Journal of Medicine, **339(7)**, 451(1998)
6. E. Tel, Ş. Okuducu, M.H. Bölükdemir, G. Tanır, International Journal of Modern Physics E **17 (3)**, 567 (2008)
7. Kohler H.S., Nucl. Phys. A, **258 (2)** 301(1976)
8. Skyrme T.H.R., Nuclear Physics **9(4)**, 615 (1958-1959)
9. H. Sahan, E. Tel, M. Sahan, A. Aydın, I.H. Sarpün, A. Kara, M. Döner, Epj Web of Conferences **100**, 01007 (2015)
10. H.M.M. Mansour, Acta Physica Polonica, **B 21**, 9 (1990)
11. E.G. Nadjakov, K.P. Marinova, Y.P. Gangrsky, Atomic Data and Nuclear Data Tables, **56 (1)** 133 (1994).
12. L.G. Qiang, J. Phys. G 17, 1 (1991).
13. E. Chabanat, P. Bonche, P. Haensel, J. Meyer, R. Schaeffer, Nucl. Phys. A **627 (4)** 710 (1997)
14. E. Chabanat, P. Bonche, P. Haensel, J. Meyer, R. Schaeffer, *Nucl. Phys. A* **635 (1-2)** 231(1998)
15. http://phys.lsu.edu/graceland/faculty/cjohnson/skhafo.f
16. EXFOR, https://www-nds.iaea.org/exfor/exfor.htm
17. U. Garuska, J. Dresler, H. Malecki, Inst. Badan Jadr. (Nucl. Res.), Swierk+Warsaw, Repts No. 1871/I/PL/A, 15 (1980)
18. A.A. Filatenkov, S.V. Chuvaev, V.A. Yakovlev, A.V. Malyshenkov, S. K. Vasil`ev, Khlopin Radiev. Inst., Leningrad Reports No.252 (1999)
19. A. Ercan, M.N. Erduran, M. Subasi, E. Gueltekin, G. Tarcan, A. Baykal, M. Bostan, Conf. on Nucl. Data for Sci. and Technol., 376 (1991)
20. Y. Ikeda, C. Konno, K. Oishi, T. Nakamura, H. Miyade, K. Kawade, H. Yamamoto, T. Katoh, JAERI Reports No.1312 (1988)
21. B. Anders, B.M. Bahal, R. Pepelnik, Ges. Kernen.-Verwertung, Schiffbau and Schiffahrt **85** 24 (1985)
22. M. Bormann, H. Jeremie, G. Andersson-Lindstroem, H. Neuert, H. Pollehn, Zeitschrift fuer Naturforschung Section A **15** 200 (1960)
23. J. Janczyszyn, L. Gorski, Journal of Radio Analytical Chemistry **14** 201 (1973)

 DOI: 10.1051/epjconf/201715401003

Proton bombarded reactions of Calcium target nuclei

Eyyup Tel[1], *Muhittin* Sahan[1], *Ismail Hakki* Sarpün[2], *Yusuf* Kavun[3], *Ali Armagan* Gök[1], *Mesut* Depedelen[1]

[1]Osmaniye Korkut Ata University, Department of Physics, Osmaniye, Turkey
[2]Afyon Kocatepe University, Physics Department, Afyonkarahisar, Turkey
[3]Manisa Celal Bayar University, Department of Physics, Manisa, Turkey

Abstract: In this study, proton bombarded nuclear reactions calculations of Calcium target nuclei have been investigated in the incident proton energy range of 1–50 MeV. The excitation functions for ^{40}Ca target nuclei reactions have been calculated by using PCROSS nuclear reaction calculation code. Weisskopf-Ewing and the full exciton models were used for equilibrium and for pre-equilibrium calculations, respectively. The excitation functions for ^{40}Ca target nuclei reactions (p,α), (p,n), (p,p) have been calculated using the semi-empirical formula Tel et al. [5].

1 Introduction

Calcium (Ca) is an important component for the living things and is necessary for human life. Free calcium metal is too reactive to occur in nature and is produced in supernova nucleosynthesis. In supernova explosions, calcium is formed from the reaction of carbon with various numbers of alpha particles (He nuclei), until the most common calcium isotope (containing 10 He nuclei) has been synthesize.

Ca with atomic number of 20 is an alkaline earth metal and is the most plentiful mineral found in the human body. While the teeth and bones especially contain the most calcium, nerve cells, body tissues, blood, and other body fluids contain the rest of the Ca. Ca is also an essential trace element in living organisms.

The Ca has five stable isotopes (^{40}Ca, ^{42}Ca, ^{43}Ca, ^{44}Ca and ^{46}Ca) and one more (^{48}Ca) that has such a long half-life. In stellar studies, one of isotopes ^{41}Ca is very important attention due to ^{41}Ca decays to ^{41}K, a critical indicator of solar-system anomalies. The 20% range in relative mass among naturally occurring calcium isotopes is greater than for any element other than H and He [1-3].

2 Calculation Model and Method

The PCROSS code uses a unified model based on the solution of the master equation in the form proposed by Cline [6] and Ribansky [7]. Griffin [8] exciton model calculations are used for $n_0 = 1$ (where $n_0 = 1$ is initial exciton number), thus are taken into account direct gamma emission;

$$n_{eq} = \sqrt{1.4gE} \quad (1)$$

where n_{eq} is equilibrium exciton number, g is single particle level density, and E is excitation energy of the compound nucleus.

A number of semi-empirical cross section have been proposed for charged and uncharged particles at the different energies. Tel et al. [5] developed the new semi-empirical cross section formulae of reactions induced by proton as follows [6-12];

$$\sigma_{(p,\alpha)} = C\sigma_{p-ne}e^{as} \quad (2)$$

σ_{p-ne} is the proton non-elastic cross section and the C and a coefficients are given in Tel et al. [5] for the (p,α) reaction at 17.9 MeV energy. r_0 is the radius of the nucleus and given as r_0=1.2 fm. Coulomb effects for proton induced reactions;

$$\sigma_{Coul} = \frac{Z^2}{(A^{1/3}+1)} \quad (3)$$

where σ_{Coul} is the proton Coulomb effect cross section.

Tel et al. [5] suggested that the empirical cross section formulae including proton inelastic cross section [13-16] and Coulomb effects of reactions induced by proton can be expressed as follows,

$$\sigma_{(p,\alpha)} = C\,\sigma_{p-ne}\sigma_{Coul}\exp(as) \quad (4)$$

where σ_{p-ne} *and* σ_{Coul} (in unit of mb) are the proton non-elastic and Coulomb effect cross section respectively [17-19]. The coefficients C and a have been determined from only empirical formulae. Z is the proton number of target nuclei.

3 Results and Discussions

In this study, proton bombarded nuclear reactions calculations of Calcium target nuclei have been investigated in the incident proton energy range of 1–50 MeV. The excitation functions for ^{40}Ca target nuclei reactions (p,α), (p,n), (p,p) have been calculated using the semi-empirical formula Tel et al. [5].

The inverse cross sections were obtained by using the optical potential parameters for neutrons proposed by Wilmore et al. [20], for protons proposed by Bechetti et al. [21], and for alpha particles by Huizenga et al. [22]. Full exciton model calculations using with PCROSS code [23] use the initial exciton number as n_o=1 of 1 neutron and 0 hole. The effects of direct reactions were considered in our calculations.

Theoretical calculations of equilibrium and pre-equilibrium for ^{40}Ca(p,α) reactions are seen starting from the same point and showed a similar upward trend, the theoretical equilibrium calculations are just a little higher than pre-equilibrium calculations as seen in Fig. 1. Theoretical calculations of ^{40}Ca(p,n) show a similar upward trend from around 17 MeV however the theoretical pre-equilibrium calculations are tends to increase after around 20 MeV according to the equilibrium calculations as seen in Fig. 2.

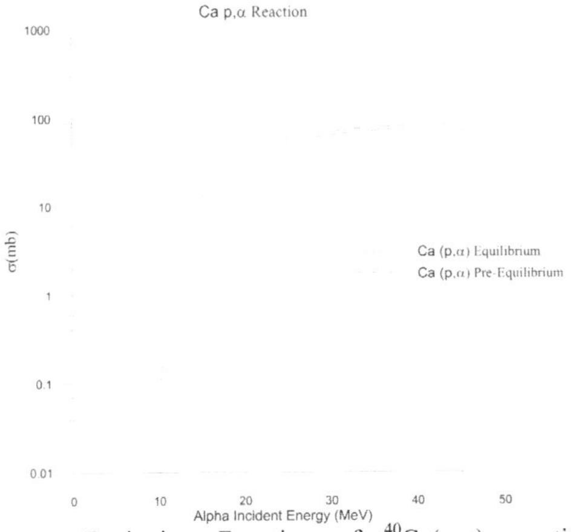

Fig.1. Excitation Function of ^{40}Ca(p,α) reactions incident proton energy 1-50 MeV.

Theoretical calculations of equilibrium and pre-equilibrium for ^{40}Ca(p,p) reaction are seen acquiring around 5 MeV but pre-equilibrium calculations have lower values according to the equilibrium calculations and they have almost the same values and show a similar tendency after 10 MeV.

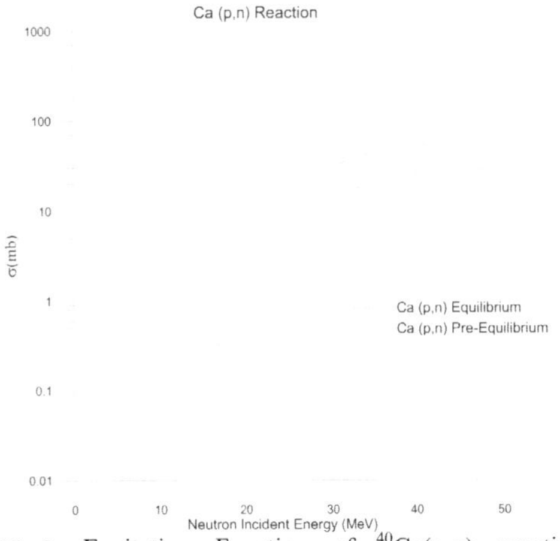

Fig.2. Excitation Function of ^{40}Ca(p,n) reactions incident proton energy 1-50 MeV.

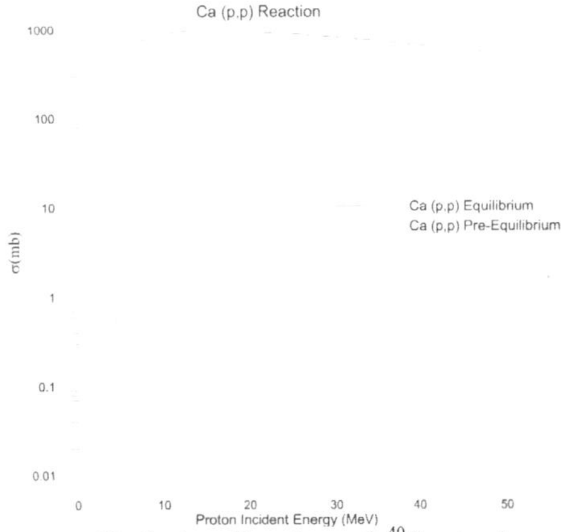

Fig.3. Excitation Function of ^{40}Ca (p,p) reactions incident proton energy 1-50 MeV.

Acknowledgment

The study was supported by Osmaniye Korkut Ata University (OKU) Science Research Projects Coordination Unit with the grant number: OKÜBAP-2017-PT3-023.

References

1. V.Y. Butko, X. Chi, D.V. Lang, and A.P. Ramirez, Appl. Phys. Lett. https://doi.org/10.1063/1.1631736 83, 4773 (2003)
2. Y.Kavun "Determination of Sr/Ca ratios in teeth samples by photoactivation analysis", Ph.D thesis Manisa Celal Bayar University, 103p. (2017)
3. Y. Kavun et al. "Preliminary results of Sr/Ca Ratio study of Teeth Samples by Photoactivation Analysis" EPJ Web of Conferences 128, 02001 (2016)
4. E. Tel, H. M. Sahin, A. Kaplan, A. Aydin, T. Altinok, Ann. Nucl. Energy **35** (2) 220, (2007)
5. E. Tel, A. Aydin, E.G. Aydın, A. Kaplan, O. Yavas, I.A. Reyhancan, PRAMANA Indian Academy of Sciences 74, 6, 931 (2010)
6. C. K. Cline, M. Blann: Nucl. Phys. A **172** 225 (1971)
7. I. Ribansky et al.: Nucl. Phys. A **205** 545. (1973)
8. J. J. Griffin, Phys. Rev. Lett. **17** 478 (1966)

9. I. Angeli, Table of nuclear root mean square charge radii, IAEANDS-163, Institute of experimental Physics, Kossuth University, H-4001 Debrecen, Pf.105, Hungary, (June 1999)
10. R.J. Harris, W.B. Shuler, M. Eckhause, R.T. Siegel, R.E. Welsh, Phys. Rev. Lett. **20** 505. (1968)
11. J.M. Akkermans et al., Phys. Rev. C **22** 73. (1980)
12. M. Belgaid, M. Asghar, Appl. Radiat. Isot. **49** 1497 (1998)
13. M. Yigit, E. Tel, Kerntechnik **79** (2014) 488.
14. M. Yigit, E. Tel, Nucl. Eng. Des. **280** 37 (2014)
15. S. Akca, E. Tel, S. Gungor, Acta Phys. Pol. A 128 (2015)
16. M. Sahan, E. Tel, H. Sahan, A. Kara, A. Aydin, A. Kaplan, I.H. Sarpun, B. Demir, S. Akca, E. Yildiz, J Fusion Energ **34**,493 (2015)
17. V.F. Weisskopf, D.H. Ewing, Phys. Rev. **57** 472 (1940)
18. W. Dilg et al., Nucl. Phys. A **217** 269. (1973)
19. F.C. Williams, Nucl. Phys. A **166** 231 (1971)
20. D. Wilmore, P.E. Hodgson, Nucl. Phys. **55** 673 (1964)
21. F.D. Becchetti, G.W. Greenlees, Phys. Rev. **182** 1190 (1969)
22. J.R. Huizenga, G. Igo, Nucl. Phys. **29** 462 (1962)
23. R. Capote et al., Final Report on Research Contract 5472/RB, INDC(CUB)-004 (Higher Institute of Nuclear Science and Technology, Cuba). Translated by the IAEA on March 1991 (PCROSS program code) (1991)

Calculations of neutron-induced alpha emission double-differential cross section of Flourine at 14.2 MeV

Muhittin Sahan, *Halide* Sahan*, *Eyyup* Tel, *Ahmet* Bulbul

Osmaniye Korkut Ata University, Department of Physics, Osmaniye, Turkey

Abstract. In this preset study, calculations of neutron-induced alpha particle emission double-differential cross section of fluorine (^{19}F) at 14.2 MeV have been calculated by using ALICE and EMPIRE model programs for six different emission angles ranging from 30° to 150°. Calculated results from the Hybrid Monte Carlo pre-equilibrium emission and the full featured Hauser-Feshbach model have been compared with the experimental (EXFOR). The calculated double-differential cross section results using three codes are in good agreement with experimental data.

1 Introduction

Light nuclides (such as FLiBe) are candidate material as an advanced blanket design in fusion reactor. Because these nuclides are usually a complex structure in the nuclear reactions around 14-15MeV neutrons. FLiBe is a molten salt made from a mixture of LiF, BeF, ThF, and UF [1]. Molten salt reactors are a good neutron moderator characterized by the use of fluoride salts as reactor coolant [1,2]. Florine and its compounds have been used widely for the nuclear applications such as processing nuclear fuel [3].

Double differential cross sections are very important for the neutronics design of fusion reactors. Double differential cross sections of charged particles emission reactions excited by 14 MeV incident neutrons is important for evaluation of nuclear heating and material damages in fusion power reactor. Theoretical calculations of double differential cross sections are very important to understand the reaction mechanism and to compare with available experimental data [4-9].

In this work, the theoretical double differential cross sections of alpha particle emission spectra at 30, 45, 60, 90, 110 and 150 degrees have been calculated for an incident neutron energy of 14.2 MeV on $^{19}F(n,xa)$ reactions. In calculations, three calculation models including EMPIRE-3.2 [10], and HMS-ALICE-2011 [11] were used. These theoretical calculations were done using Hauser-Feshbach statistical model [12], statistical model with compound, pre-equilibrium and direct [13].

2 Results and Discussions

In this study, double differential cross section emission spectra for the $^{19}F(n,xa)$ fusion materials have been calculated by using the Hybrid Monte Carlo (HMS) for pre-equilibrium emissions at 14.2 MeV incident neutron energy and emission angles ranging from 30° to 150° (30°, 45°, 60°, 90°, 110° and 150°). In calculations, ALICE and EMPIRE computer codes were used. The calculated values of the emitted alpha particles at given angles were compared with experimental data.

The calculated double differential cross section results for $^{19}F(n,xa)$ reaction were given in Figs. 1-6, respectively. The experimental data for $^{19}F(n,xa)$ in literature are taken from Kondo et al. [14], Terada et al. [15] with using the EXFOR library [16].

Theoretical double differential cross section calculations at 30° for $^{19}F(n,xa)$ were given in Fig. 1. In Fig 1, we compared our double differential cross section calculations with the experimental data of Kondo et al. [14]. Theoretical double differential cross section calculations from ALICE code are close agreement with experimental data by Kondo et al. [14] especially at the continuum from 4 to 11 MeV while EMPIRE results are similar with experimental data between 4 and 7 MeV.

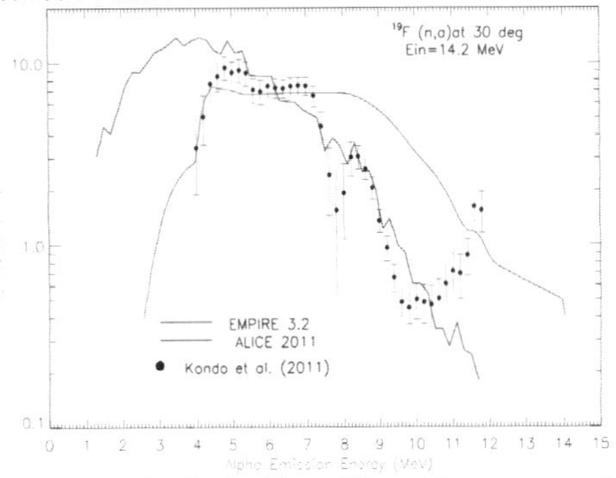

Fig.1. Calculated double differential cross section calculations of alpha particle at 30°.

* Corresponding author: halidesahan@osmaniye.edu.tr

Fig. 2 shows two double differential cross section calculations for the alpha emission spectrum at 45° for an incident neutron energy of 14.2 MeV on ^{19}F(n,xa) reaction. Experimental data from Terada et al. [14] are in general agreement with ALICE results.

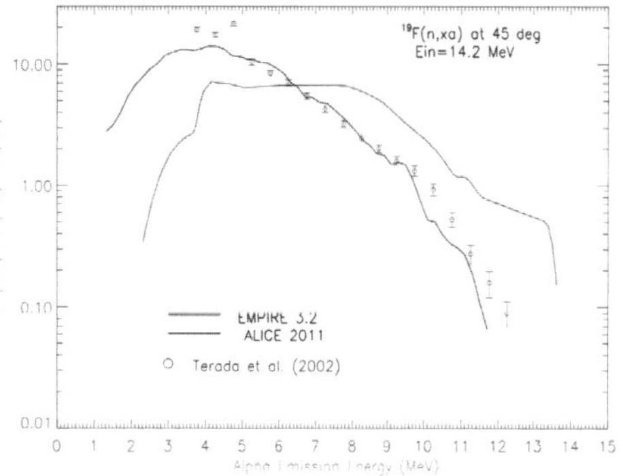

Fig. 2. The same with Fig. 1, but for 45°.

The calculated double differential cross section alpha emission spectra comparing with the experimental data by Terada et al. [15] at emission angle of 60° and for 14.2 MeV incident neutron energy are given in Fig. 3. Similar to the 45°, ALICE results are in general agreement with experimental data especially between 4 MeV and 12 MeV.

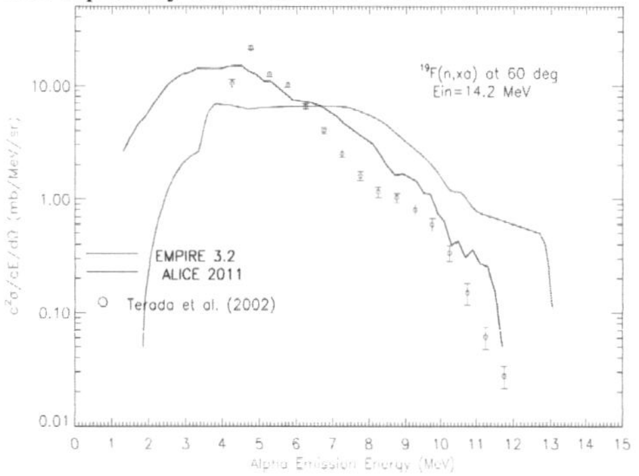

Fig. 3. The same with Fig. 1, but for 60°.

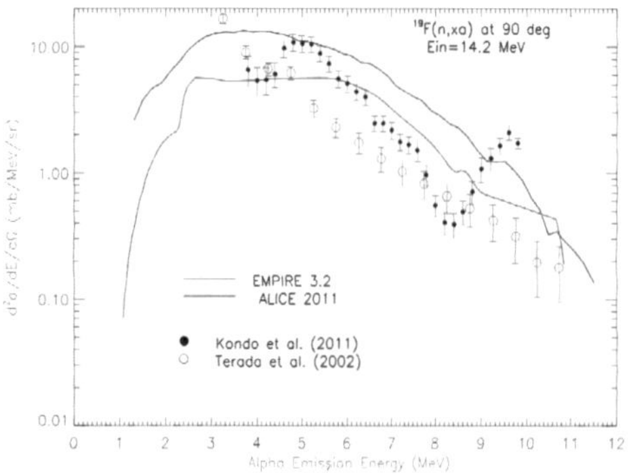

Fig. 4. The same with Fig. 1, but for 110°.

Fig. 4 shows the results of double differential cross section for alpha emitted in reactions of incident neutron energy of14.2 MeV at emission angle of 90° degree for ^{19}F(n,xa) reaction. In Fig. 4, the calculated data for 14.2 MeV neutrons are compared with measurement data by Terada et al. and Kondo et al. [14]. Both theoretical cross section data calculated by EMPIRE-3.2 and ALICE codes from 3 MeV to 11 MeV are approximately close agreement with experimental data.

Fig.5 shows the double differential alpha emission cross sections of fluorine for 14.2 MeV neutrons at angles of 110 degree in comparison with experimental data measured by Terada et al. [15] and Kondo et al. [14]. From Fig. 5, it is seen that there are two peaks of experimental data given by Kondo et al. [14] at 5MeV and 9 MeV. Double differential cross section curves of ALICE and EMPIRE are almost constant throughout the entire alpha emission energy range. The comparison of calculated data from three codes at this incident energy for 110° is in good agreement with experimental data.

Fig. 6 shows the results of double differential cross section for alpha emitted in reactions of an incident neutron energy 14.2 MeV at 150° for ^{19}F(n,xa) reaction. Fig. 6 also shows the comparisons of double differential cross sections among the calculated data and available experimental data by Kondo et al. [14]. Comparing theoretical calculations from ALICE and EMPIRE codes with experimental data, it is seen that experimental data is lower cross section values between 4.5 MeV and 7.5 MeV.

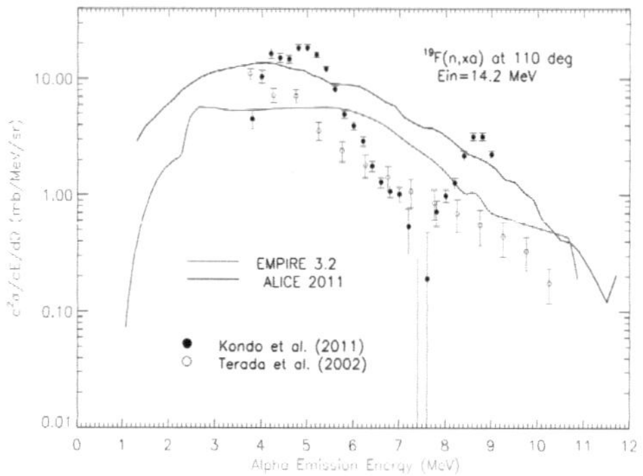

Fig. 5. The same with Fig. 1, but for 90°.

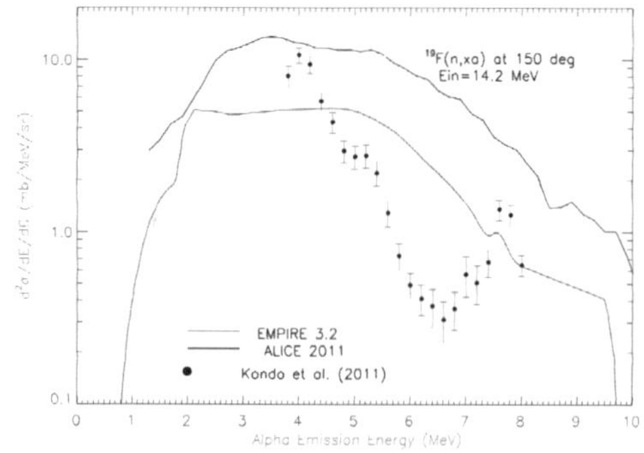

Fig. 6. The same with Fig. 1, but for 150°.

We calculated theoretical energy differential cross section for $^{19}F(n,x\alpha)$ reaction as seen from Fig. 7. The energy differential cross sections are plotted versus alpha emission energy at 14.2 MeV in Fig. 7. As seen from Fig. 7, total, direct, pre-equilibrium, and compound energy differential cross sections are given. Moreover, experimental energy differential cross section obtained from Terada et al. [15] is plotted to compared with theoretical results. Calculations of dσ/dE predicts that probability of alpha energy are generally emitted from compound nucleus mechanism.

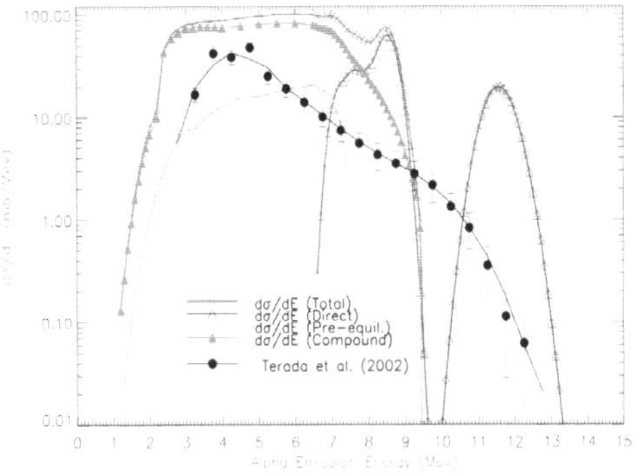

Fig. 7. Theoretical differential energy spectra of emitted alphas from $^{19}F(n,x\alpha)$ reaction at 14.2 MeV neutrons.

We also obtained angular distributions of the double differential alpha emission cross sections at the incident energies of 0-2, 2-4, 4-6, 6-8, 8-10, and 10-12 MeV at incident energy of 14.2 MeV and then we plotted logarithmic energy values versus angle values from 0 to 180° as seen Fig. 8.

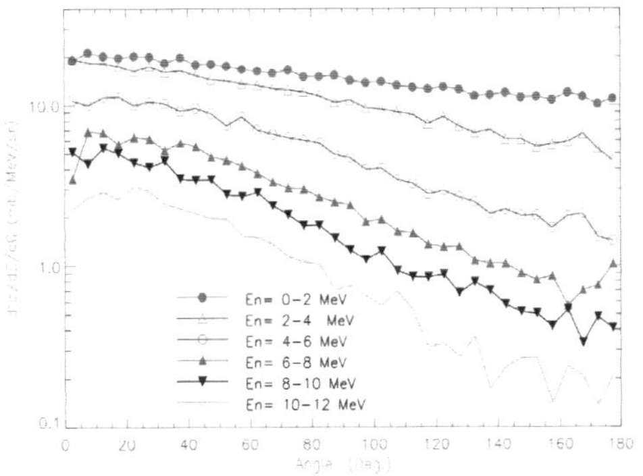

Fig. 8. Angular distributions of emission alpha from from $^{19}F(n,x\alpha)$ reaction for different energies at incident energy of 14.2 MeV

3 Conclusion

In present work, we used ALICE, EMPIRE model codes to calculate double differential cross sections spectra for the $^{19}F(n,x\alpha)$ structural fusion materials at 14.2 MeV for different emission angles ranging from 30 to 150 degrees.

Conclusions are briefly summarized as;

- The model calculations provide the good description of the shape and magnitude of double differential cross of alpha emission for selected emission angles except for 150 deg.
- It is clearly seen that there are very good agreements in the energy region less than about alpha emission energy 11 MeV at all emission angles.
- For all angles between 30-150 degrees, alpha emission energy values are seen up to maximum 12 MeV and hence these energy range are decrease while emission angles increase.
- It is also clearly seen that there are very good agreements in the full alpha energy range for all angles.
- For better double differential cross alpha emission further theoretical studies are required.

References

1. R.W. Moir and E. Teller, Nuclear Technology, **151**, 334 (2005)
2. L. Mei, X. Cai, D. Jiang, J. Chen, Y. Zhu, Y. Liu, and X. Wang, J. of Nuclear Science and Technology, **50,** 682 (2013)
3. K. Kondo, I. Murata, K. Ochiai, N. Kubota, H. Miyamaru, C. Konno, and T. Nishitani, J. of Nuclear Science and Technology, **48**, 1146 (2011)
4. I.H. Sarpun, A. Aydin, A. Koning, J. Fusion Energ. **35**, 725 (2016)
5. H. Şahan, E. Tel, M. Yiğit, J. Fusion Energ. **34**, 16 (2015)
6. M. Sahan, E. Tel, H. Sahan, A. Kara, A. Aydin, A. Kaplan, I.H. Sarpun, B. Demir, S. Akca, E. Yildiz, J. of Fusion Energy, **34,** 493 (2015)
7. A. Sagara, H. Yamanishi, S. Imagawa, et al., Fusion Eng. Des., **49**, 661 (2000).
8. K. Kondo, S. Takagi, I. Murata, and H. Miyamam, JAERI-Conf., 003 (2005)
9. Y. Terada, H. Takagi, Kokooo, I. Murata, and A. Takahashi, Journal of Nuclear Science and Technology, **2**, 413 (2002)
10. M. Herman, et al., EMPIRE-3.2 Malta modular system for nuclear reaction calculations and nuclear data evaluation, User's Manual (2013)
11. C.H.M. Broeders, A. Yu. Konobeyev, Yu. A. Korovin, V.P. Lunev, and M. Blann, ALICE/ ASH Manual, FZK 7183 (2006)
12. W. Hauser, H. Feshbach, Phys. Rev. C, **87**, 366 (1952)
13. A.J. Koning, J.P. Delaroche, Nucl. Phys., A **713**, 231 (2003)
14. K. Kondo, I. Murata, K. Ochiai, N. Kubota, H. Miyamaru, C. Konno and T. Nishitani, Journal of Nuclear Science and Technology, **48**, 1146 (2011)
15. Y. Terada, H. Takagi, Kokooo, I. Murata, and A. Takahashi, Journal of Nuclear Science and Technology, Supplement, **2**, 413 (2002)
16. http://www.nndc.bnl.gov/exfor/exfor00.htm

DOI: 10.1051/epjconf/201715401005

Gamma ray heating rates due to chromium isotopes in stellar core during late stages of high mass stars (>10M$_\odot$)

Jameel-Un Nabi and *Muhammad* Fayaz

GIK Institute of Engineering Sciences and Technology, Topi 23640, Khyber Pakhtunkhwa, Pakistan

Abstract. Gamma ray heating rates are thought to play a crucial role during the pre-supernova stage of high mass stars. Gamma ray heating rates, due to $\beta^{\pm}$ -decay and electron (positron) capture on chromium isotopes, are calculated using proton-neutron quasiparticle random phase approximation theory. The electron capture significantly affects the lepton fraction (Y_e) and accelerates the core contraction. The gamma rays emitted as a result of weak processes heat the core and tend to hinder the cooling and contraction due to electron capture and neutrino emission. The emitted gamma rays tend to produce enormous entropy and set the convection to play its role at this stage. The gamma heating rates, on $^{50\text{-}60}$Cr, are calculated for the density range $10 < \rho$ (g.cm^{-3}) $< 10^{11}$ and temperature range $10^7 < T$ (K) $< 3.0\times10^{10}$.

1 Introduction

Stars with masses greater than 10M$_\odot$ proceed beyond oxygen burning and develop an electron degenerate iron core [1]. The electron degeneracy pressure supports the core against the gravity. The silicon burning in shell keeps growing the core towards the Chandrasekhar mass limit (~1.4 M$_\odot$). At this stage electron degeneracy pressure cannot combat the enormous self gravity. The core becomes unstable and begins to collapse with a quarter of the speed of light [2]. The hot stellar core generates energetic gamma rays to initiate photodisintegration, which causes a complete collapse of the core. This way the gamma rays can remarkably affect the core growth, core collapse and core bounce during the last evolution phases of the stars. Also, the gamma heating rates cause convection to bring fresh material to the core and to continue the fusion process.

Weak interaction processes in massive stars provide the energy budget during the final stages of stellar life. Ref. [3] argues that neutrino cooling remarkably reduces the heating effect of gamma rays. Owing to core mass and its composition, these reactions can release 10 times more heat at densities ~10^{11} g.cm^{-3}. The electron capture on various nuclei is sensitive to temperature [4]; so is the associated gamma heating. It is, therefore, worthwhile to calculate the associated gamma heating rates for the iron peak nuclei. During these processes, the stellar core stiffens to nuclear density and the collapsing core bounces against the nuclear matter. The parts of heavy-element core and outer shells are ejected into the interstellar medium [5]. The gamma heating, at this stage, may also contribute to revive the stalled shock, leading to explosive burning and convection.

The electron capture and β-decay rates, and associated neutrino-cooling and gamma heating rates, are sensitive to Gamow-Teller (GT) strength distributions. The neutrino-cooling occurs due to the production of (anti)neutrinos as a result of capture and decay processes. On the other hand, gamma ray heating results as the excited daughter nuclei (of capture and decay processes) return to their ground states. The GT transitions, in *fp*-shell nuclei, are crucially important in pre-supernova phases and the core collapse [6-9].

Isotopes of chromium play an important part in the pre-supernova evolution of massive stars. The measured data of GT strength in chromium isotopes is rather scarce. The decay of ^{46}Cr was studied by Ref. [10], by producing ^{46}Cr via the reaction ^{32}S (^{16}O, 2*n*). Later Ref. [11] observed the *β*-decay of ^{46}Cr to the 1_1^+ state at 993 keV excitation energy in ^{46}V at RIKEN. Ref. [12] did a ^{50}Cr(^{3}He,t)^{50}Mn measurement up to 5 MeV in daughter. A high resolution ^{50}Cr(^{3}He,t)^{50}Mn measurement at an incident energy of 140 MeV/nucleon and at 0^0 was performed by Ref. [13] at RCNP, Japan. The authors measured $B(GT)^-$ strength up to 12 MeV in ^{50}Mn. The simulation of astrophysical events (e.g. core-collapse supernovae), requires GT strength distribution for hundreds of nuclei. Therefore, theoretical estimates for GT strength distributions, microscopic in nature, are very much in demand.

The rate of change of Y_e (lepton-to-baryon fraction) during pre-supernova evolution is one of the key inputs to generate a successful explosion. Ref. [14] found that the nuclei $^{51-58}$Cr, for electron capture, and $^{53\text{-}57,59,60}$Cr for *β*-decay are of astrophysical significance in controlling Y_e within the range $0.40 \le Y_e \le 0.50$. Ref. [15] found that ^{50,51,53}Cr were crucial for decreasing the Y_e and $^{53\text{-}56}$Cr played a significant role in increasing the Y_e content of the stellar matter.

These studies motivated us to calculate the GT strength distributions and the associated gamma heating rates due to isotopes of chromium ($^{50-60}$Cr), both in electron capture and β-decay directions. The calculated GT distributions are compared with previous works for few even-even cases. We used the proton neutron quasiparticle random phase approximation (pn-QRPA) model to calculate the GT strength distributions and associated gamma heating rates in stellar matter. The formalism of our calculation is described briefly in the next section. We compare our results with other model calculations and measurements in Section 3.

2 Model Description and Formalism

A brief description of the pn-QRPA model is presented. For the detailed model formalism, we refer to Ref. [16]. The Hamiltonian of the pn-QRPA model is given by

$$H^{QRPA} = H^{SP} + V^{pair} + V^{ph}_{GT} + V^{pp}_{GT}. \qquad (1)$$

Wave functions and single particle energies were calculated using the Nilsson model. Pairing in nuclei was treated within the BCS approximation. The proton-neutron residual interaction occurs through particle-hole (ph) and particle-particle (pp) channels. In our pn-QRPA model, these interactions were given a separable form as V^{ph}_{GT} for the particle-hole Gamow-Teller force and V^{pp}_{GT} for the particle-particle Gamow-Teller force. Our calculated GT strength distributions fulfilled the model-independent Ikeda sum rule [17].

The other requisite model parameters are the Nilsson potential parameters, the pairing gaps, the nuclear quadrupole deformations, and the Q-values of the reactions. Nilsson potential parameters were adopted from Ref. [18] and the Nilsson oscillator constant was chosen as $\omega = 41A^{-1/3}MeV$. The relation for pairing gaps was taken as: $\Delta_p = \Delta_n = 12/\sqrt{A}\, MeV$.

The nuclear quadrupole deformation parameters "β" were taken from Ref. [19]. For even-even nuclei it was calculated using:

$$\beta = \frac{4\pi}{3ZR_0^2}[B(E2)\uparrow /e^2]^{1/2}, \qquad (2)$$

where $B(E2)\uparrow$ is the reduced electric quadrupole transition probability, Z is the atomic number, and $R_0^2 = 0.0144A^{2/3}$ barn. For remaining nuclei the deformation was taken from Ref. [20]. Q-values were taken from Ref. [21]. We used a model space of 7 ω in our model calculation.

The basic formalism for calculation of stellar weak rates was adopted from the pioneering work of Fuller, Fowler and Newman (FFN) [6-9]. However, unlike the work of FFN, we calculated GT strength distributions from all parent excited states in a microscopic way. The weak decay rates of a transition from the ith parent to the jth daughter state is given by :

$$\lambda_{ij} = (\frac{\ln 2}{D})[f_{ij}(T,\rho,E_f)][B(F)_{ij} + (g_A/g_V)^2 B(GT)_{ij}]. \qquad (3)$$

The value of D = 6143s was taken from Ref. [22]. *B(F)ij* denote the total reduced transition probability due to Fermi interaction and *B(GT)ij* due to Gamow-Teller interactions. The value of $(g_A/g_V) = -1.2694$ (in Eq.3), was taken from [23]. The f_{ij} are the phase space integrals which were calculated (in natural units $\hbar = m_e = c = 1$) for electron (upper signs) or positron (lower signs) emissions, using the relation:

$$f_{ij} = \int_1^{w_m} w\sqrt{w^2-1}(w_m - w)^2 F(\pm Z, w)(1-G\ )dw, \qquad (4)$$

the phase space for continuum positron *(lower signs)* or electron *(upper signs)* capture was calculated using:

$$f_{ij} = \int_{w_l}^{\infty} w\sqrt{w^2-1}(w_m + w)^2 F(\pm Z, w)G\ dw. \qquad (5)$$

In Eqs. (4) and (5) "w" is the total energy (rest+kinetic) of the electron or positron, "w_l" is the total threshold energy (rest+kinetic) for positron (or electron) capture and "w_m" is the total β-decay energy. $F(\pm Z,w)$ are the Fermi functions and $G_\pm$ is the Fermi-Dirac distribution function for positrons (electrons).

The ratio of electron to baryon is given as :

$$\rho Y_e = \frac{1}{\pi^2 N_A}(\frac{m_e c}{\hbar})^3 \int_0^{\infty} (G_- - G_+)p^2 dp, \qquad (6)$$

where $p = (w^2-1)^{1/2}$ is the electron or positron momentum, and Eq. (6) has the units of *mole-cm*$^{-3}$. This equation was used for an iterative calculation of Fermi energies for selected values of ρY_e and T.

The total gamma ray heating rate was calculated using the relation:

$$\lambda^{\gamma} = \sum_{ij} P_i \lambda_{ij} E_j, \qquad (7)$$

where "P_i" is the probability of occupation of parent excited states and obey the Boltzmann distribution, λ_{ij} gives the sum of the electron(positron) capture and positron(beta) decay rates, $\lambda_{ij} = \lambda_{ij}^{ec(pc)} + \lambda_{ij}^{pd(bd)}$, and E_j is the energy of daughter's excited state.

3 Results and Discussion

In this section, we show our calculation of GT strength distributions and gamma heating rates for few even-even isotopes of chromium. The calculated GT strength in the *β-decay* direction for ^{50}Cr is shown in Fig. 1. The upper two panels show the measured GT distributions for ^{50}Cr. The panel (Exp.1) shows the measurement of (^{3}He, t) experiment up to 5 MeV by Ref. [12]. The lower panel (Exp. 2) shows the B(GT)- distribution up to 12 MeV in daughter measured by Ref. [13]. The bottom panel of Fig. 1 further compares the pn-QRPA calculated GT strength distribution with the shell model calculation using the KB3G interaction, by Ref. [24]. It is noted that pn-QRPA calculated transitions are placed at lower energies and of bigger magnitude in daughter than the shell model results. It is inferred that the calculated GT distribution is in good conciliation with Exp. 2. The pn-QRPA calculated GT- distribution in ^{52}Cr is compared with shell model calculation in Fig. 2. The shell model results spread over an energy range of 5-15 MeV in daughter. While the pn-QRPA places the energy centroid at low excitation energy of 5.41 MeV in ^{52}Mn. Fig. 3 shows a similar comparison of our GT- strength in ^{54}Cr with shell model results. The GT strength is placed at energy intervals of 2.5-10 MeV in pn-QRPA model and 8-16 MeV in shell model calculation. The pn-QRPA calculated ground and excited states GT strength distributions for $^{50\text{-}60}$Cr are available and may be requested as ASCII files from the corresponding author.

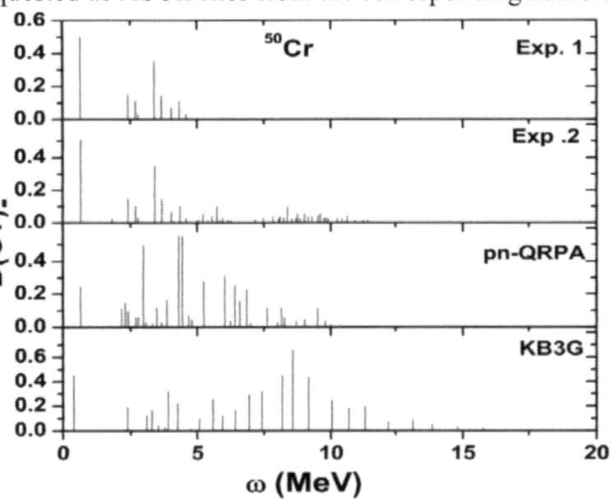

Figure 1 Comparison of B(GT)- strength distributions in ^{50}Cr with measured and calculated data. Exp. 1 shows measured distribution by [12], Exp. 2 by [13] while KB3G shows shell model calculation by [24]. ω represents excitation energy in ^{50}Mn in units of MeV.

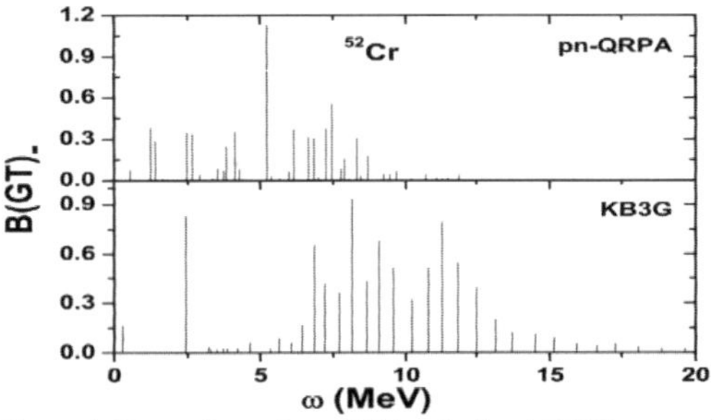

Figure 2 Comparison of pn-QRPA calculated B(GT)- strength distributions in ^{52}Cr with shell model calculation [24]. ω represents excitation energy in ^{52}Mn in units of MeV.

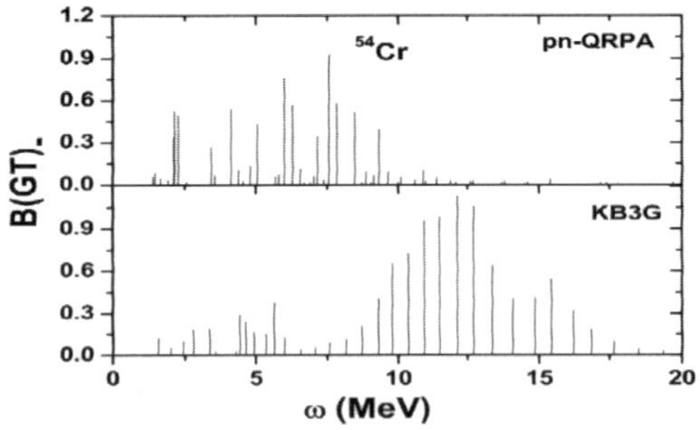

Figure 3 Comparison of pn-QRPA calculated B(GT)- strength distribution in ^{54}Cr with shell model calculation [24]. ω represents excitation energy in ^{54}Mn in units of MeV.

The gamma ray heating rates due to ^{50}Cr, ^{52}Cr, and ^{54}Cr for selected densities (1st column) and temperatures (2nd column) are shown in Table 1. The value 1.00×10^{-100} MeV.s^{-1} means that the rates are negligibly small. These rates increase with temperature but those due to electron capture increase at faster rate than those due to β-decay for densities 10 to 10^7 g.cm^{-3}. However, at 10^{11} g.cm^{-3}, the rates due to electron capture dominate over β-decay by orders of magnitudes and show nearly constant trend with temperature. In case of ^{54}Cr, the gamma heating rates at lower densities dominate due to β-decay and at higher density due to electron capture. Next, we present the gamma heating rates due to ^{56}Cr, ^{58}Cr, and ^{60}Cr in Table 2. Here again, a similar trend is witnessed for all these isotopes of chromium.

The ASCII files of calculated gamma heating rates due to isotopes of chromium, $^{50\text{-}60}$Cr, are available on a fine temperature-density grid, suitable for interpolation process, and may be requested from the corresponding author. It is hoped that the calculated heating rates may contribute to a better understanding of the dynamics of core evolution. We are in a process of calculating the neutrino cooling rates due to isotopes of chromium in stellar matter and hope to soon report our findings.

Acknowledgement: J.-U. Nabi would like to acknowledge the support of the Higher Education Commission Pakistan through the HEC Project No. 5557/KPK/NRPU/R&D/HEC/2016, Pakistan Science Foundation through the project PSF-TUBITAK/KP-GIKI (02) and TUBITAK project code 2221-1059B211700192.

References

1. R. Buras, M. Rampp, H.T. Janka, K. Kifonidis, Phys. Rev. Lett., **90**(24), 241101(2003).
2. J. José, C. Iliadis, Rep. Prog. Phys., **74**(9), 096901 (2011).
3. S. Gupta, E.F. Brown, H. Schatz, P. Möller, K.-L. Kratz, Astrophys. J. **662**(2), 1188 (2007).
4. P.G. Giannaka, T.S. Kosmas, Adv. High Energy Phys. **2015** (2015).
5. H.T. Janka, K. Langanke, A. Marek, G. Martínez-Pinedo, B. Müller, Phys. Rep., **442**(1), 38(2007).
6. G.M. Fuller, W.A. Fowler, M.J. Newman, Astrophys. J. Suppl. Ser. **42**, 447 (1980).
7. G.M. Fuller, W.A. Fowler, M.J. Newman, Astrophys. J. Suppl. Ser. **48**, 279 (1982).
8. G.M. Fuller, W.A. Fowler, M.J. Newman, Astrophys. J. **252**, 715 (1982).
9. G.M. Fuller, W.A. Fowler, M.J. Newman, Astrophys. J. **293**, 1 (1985).
10. J. Zioni, A.A. Jaffe, E. Friedman, N. Haik, R. Schectman, D. Nir, Nucl. Phys. A **181**, 465 (1972).
11. T.K. Onishi et al., Phys. Rev. C **72**, 024308 (2005).
12. Y. Fujita, B. Rubio, W. Gelletly, Prog. Part. Nucl. Phys. **66**, 549 (2011).
13. T. Adachi et al., Nucl. Phys. A **778**, 70c (2007).
14. M.B. Aufderheide, I. Fushiki, S.E. Woosley, Astrophys. J. Suppl. **91**, 389 (1994).
15. A. Heger, S.E. Woosley, G. Martínez-Pinedo, K. Langanke, Astrophys. J. **560**, 307 (2001).
16. J. -U. Nabi, H. V. Klapdor-Kleingrothaus, At. Data Nucl. Data Tables **71**, 149 (1999).
17. I. Ikeda, Prog. Theor. Phys., **31**, 434 (1964).
18. S. G. Nilsson, Mat. Fys. Medd. Dan. Vid. Selsk **29**, 16 (1955).
19. S. Raman, C. W. Nestor, P. Tikkanen, At. Data Nucl. Data Tables **78**(1), 1 (2001).
20. P. Möller A. J. Sierk, T. Ichikawa, H. Sagawa, At. Data Nucl. Data Tables **109**, 1 (2016).
21. M. Wang, G. Audi, A. H. Wapstra, F. G. Kondev, M. MacCormick, X. Xu, B. Pfeiffer, Chin. Phys. C, **36**, 1603 (2012).

22. J. C. Hardy, I. S. Towner, Phys. Rev. C **79(5)**, 055502 (2009).
23. K. Nakamura (Particle Data Group), J. Phys. G: Nucl. Part. Phys. **37**(7A), 075021 (2010).
24. I. Petermann, G. Martinez-Pinedo, K. Langanke, E. Caurier, Eur. Phys. J. A **34**, 319 (2007).

Table 1 Calculated gamma heating rates (in MeV.s^{-1}) due to weak rates on ^{50}Cr, ^{52}Cr, and ^{54}Cr at selected densities and temperatures. The term λ_{γ}^{pc+ed} is for gamma heating rates due to positron capture and β^{-} decay while λ_{γ}^{ec+pd} is for gamma heating rates due to electron capture and positron decay.

ρ(g.cm^{-3})	T(K)	^{50}Cr		^{52}Cr		^{54}Cr	
		λ_{γ}^{pc+ed}	λ_{γ}^{ec+pd}	λ_{γ}^{pc+ed}	λ_{γ}^{ec+pd}	λ_{γ}^{pc+ed}	λ_{γ}^{ec+pd}
	1.00×10^{7}	1.00×10^{-100}	1.00×10^{-100}	1.00×10^{-100}	1.00×10^{-100}	1.00×10^{-100}	1.00×10^{-100}
	1.00×10^{8}	1.00×10^{-100}	5.02×10^{-67}	1.00×10^{-100}	1.00×10^{-100}	1.00×10^{-100}	1.00×10^{-100}
1.00×10^{1}	1.00×10^{9}	1.13×10^{-44}	3.57×10^{-15}	1.38×10^{-31}	1.51×10^{-29}	1.19×10^{-15}	4.33×10^{-43}
	1.00×10^{10}	2.45×10^{-4}	6.47×10^{-2}	3.98×10^{-3}	8.20×10^{-3}	5.89×10^{-2}	8.71×10^{-4}
	3.00×10^{10}	1.42×10^{1}	2.94×10^{2}	5.96×10^{1}	1.59×10^{2}	1.72×10^{2}	1.10×10^{2}
	1.00×10^{7}	1.00×10^{-100}	1.00×10^{-100}	1.00×10^{-100}	1.00×10^{-100}	1.00×10^{-100}	1.00×10^{-100}
	1.00×10^{8}	1.00×10^{-100}	7.50×10^{-64}	1.00×10^{-100}	1.00×10^{-100}	1.00×10^{-100}	1.00×10^{-100}
1.00×10^{4}	1.00×10^{9}	1.01×10^{-45}	4.01×10^{-14}	1.70×10^{-32}	1.69×10^{-28}	1.06×10^{-16}	4.84×10^{-42}
	1.00×10^{10}	2.45×10^{-4}	6.47×10^{-2}	3.98×10^{-3}	8.22×10^{-3}	5.89×10^{-2}	8.71×10^{-4}
	3.00×10^{10}	1.42×10^{1}	2.95×10^{2}	5.96×10^{1}	1.59×10^{2}	1.72×10^{2}	1.10×10^{2}
	1.00×10^{7}	1.00×10^{-100}	1.00×10^{-100}	1.00×10^{-100}	1.00×10^{-100}	1.00×10^{-100}	1.00×10^{-100}
	1.00×10^{8}	1.00×10^{-100}	2.93×10^{-28}	1.00×10^{-100}	1.00×10^{-100}	1.00×10^{-100}	1.00×10^{-100}
1.00×10^{7}	1.00×10^{9}	5.83×10^{-50}	3.99×10^{-9}	1.95×10^{-35}	1.53×10^{-23}	1.21×10^{-21}	4.82×10^{-37}
	1.00×10^{10}	1.95×10^{-4}	8.09×10^{-2}	3.18×10^{-3}	1.03×10^{-2}	4.76×10^{-2}	1.09×10^{-3}
	3.00×10^{10}	1.41×10^{1}	2.97×10^{2}	5.92×10^{1}	1.60×10^{2}	1.71×10^{2}	1.11×10^{2}
	1.00×10^{7}	1.00×10^{-100}	1.37×10^{5}	1.00×10^{-100}	8.57×10^{4}	1.00×10^{-100}	1.68×10^{4}
	1.00×10^{8}	1.00×10^{-100}	1.36×10^{5}	1.00×10^{-100}	8.54×10^{4}	1.00×10^{-100}	1.68×10^{4}
1.00×10^{11}	1.00×10^{9}	1.00×10^{-100}	1.36×10^{5}	1.00×10^{-100}	8.55×10^{4}	1.00×10^{-100}	1.67×10^{4}
	1.00×10^{10}	2.40×10^{-16}	1.44×10^{5}	4.16×10^{-15}	9.68×10^{4}	1.75×10^{-13}	2.89×10^{4}
	3.00×10^{10}	1.95×10^{-3}	2.36×10^{5}	8.20×10^{-3}	1.82×10^{5}	2.40×10^{-2}	1.65×10^{5}

Table 2 Same as Table 3.1 but for ^{56}Cr, ^{58}Cr, and ^{60}Cr.

ρ(g.cm^{-3})	T(K)	^{56}Cr		^{58}Cr		^{60}Cr	
		λ_{γ}^{pc+ed}	λ_{γ}^{ec+pd}	λ_{γ}^{pc+ed}	λ_{γ}^{ec+pd}	λ_{γ}^{pc+ed}	λ_{γ}^{ec+pd}
	1.00×10^{7}	3.42×10^{-4}	1.00×10^{-100}	9.40×10^{-2}	1.00×10^{-100}	2.03×10^{0}	1.00×10^{-100}
	1.00×10^{8}	3.42×10^{-4}	1.00×10^{-100}	9.40×10^{-2}	1.00×10^{-100}	2.03×10^{0}	1.00×10^{-100}
1.00×10^{1}	1.00×10^{9}	3.42×10^{-4}	1.56×10^{-53}	9.40×10^{-2}	1.11×10^{-64}	2.03×10^{0}	1.26×10^{-75}
	1.00×10^{10}	5.20×10^{-1}	4.32×10^{-4}	1.06×10^{1}	9.23×10^{-5}	7.23×10^{1}	7.64×10^{-6}
	3.00×10^{10}	2.32×10^{2}	9.59×10^{1}	1.29×10^{3}	5.22×10^{1}	3.21×10^{3}	2.76×10^{1}
	1.00×10^{7}	3.37×10^{-4}	1.00×10^{-100}	9.35×10^{-2}	1.00×10^{-100}	2.03×10^{0}	1.00×10^{-100}
	1.00×10^{8}	3.38×10^{-4}	1.00×10^{-100}	9.38×10^{-2}	1.00×10^{-100}	2.03×10^{0}	1.00×10^{-100}
1.00×10^{4}	1.00×10^{9}	3.41×10^{-4}	1.75×10^{-52}	9.38×10^{-2}	1.24×10^{-63}	2.03×10^{0}	1.26×10^{-75}
	1.00×10^{10}	5.20×10^{-1}	4.32×10^{-4}	1.06×10^{1}	9.23×10^{-5}	7.23×10^{1}	1.66×10^{-6}
	3.00×10^{10}	2.32×10^{2}	9.59×10^{1}	1.30×10^{3}	5.22×10^{1}	3.22×10^{3}	2.76×10^{1}
	1.00×10^{7}	4.84×10^{-5}	1.00×10^{-100}	6.44×10^{-2}	1.00×10^{-100}	1.65×10^{0}	1.00×10^{-100}
	1.00×10^{8}	4.84×10^{-5}	1.00×10^{-100}	6.43×10^{-2}	1.00×10^{-100}	1.65×10^{0}	1.00×10^{-100}
1.00×10^{7}	1.00×10^{9}	6.03×10^{-5}	1.75×10^{-47}	5.53×10^{-2}	1.24×10^{-58}	1.66×10^{0}	1.40×10^{-69}
	1.00×10^{10}	4.35×10^{-1}	5.42×10^{-4}	1.00×10^{1}	1.16×10^{-4}	7.06×10^{1}	9.59×10^{-6}
	3.00×10^{10}	2.30×10^{2}	9.68×10^{1}	1.29×10^{3}	5.27×10^{1}	3.20×10^{3}	2.79×10^{1}
	1.00×10^{7}	1.00×10^{-100}	6.55×10^{3}	1.00×10^{-100}	2.06×10^{3}	1.00×10^{-100}	4.99×10^{2}
	1.00×10^{8}	1.00×10^{-100}	6.56×10^{3}	1.00×10^{-100}	2.05×10^{3}	1.00×10^{-100}	5.01×10^{2}
1.00×10^{11}	1.00×10^{9}	1.00×10^{-100}	6.55×10^{3}	1.00×10^{-100}	2.05×10^{3}	1.23×10^{-92}	4.99×10^{2}
	1.00×10^{10}	1.00×10^{-11}	1.75×10^{4}	1.26×10^{-9}	4.98×10^{3}	6.01×10^{-8}	2.31×10^{3}
	3.00×10^{10}	3.36×10^{-2}	1.52×10^{5}	2.47×10^{-1}	8.41×10^{4}	1.52×10^{0}	5.69×10^{5}

DOI: 10.1051/epjconf/201715401006

Calculations of (n,2n) reaction cross sections for Barium isotopes from 5 to 20 MeV

Halide Sahan, *Muhittin* Sahan, *Eyyup* Tel

Osmaniye Korkut Ata University, Department of Physics, Osmaniye, Turkey

Abstract. In this study, the excitation functions of (n,2n) reactions for $^{30,32,34,35,37,38}Ba$ isotopes are calculated using TALYS 1.6, EMPIRE-3.2.2, and ALICE-GDH codes based on statistical model up to 20 MeV. Moreover, the cross section for each isotope have also been estimated at 14.2 MeV using semi empirical formula developed by four different authors. The calculated and estimated cross-sections are compared with experimental cross-sections from EXFOR and compared with the evaluation data in ENDF/B-VII.1 library. Results are close agreement with the experimental data from literature.

1 Introduction

The nuclear reaction models have been developed estimate the reaction cross-sections due to lack of experimental data [1-4]. Most of existing information concerning the mechanism of (n, 2n) have been acquired from the data taken at neutron energies near 14 MeV. Importance of theoretical cross-section calculations for fast fission and fusion reactors increase in recent years. Fission products accumulate in an operating reactor as the fuel undergoes fission [5]. Cross sections for (n,2n) reactions induced by neutrons of 14-15 MeV energies have been calculated by many researches covering all the elements of the periodic table.

In this study, the (n,2n) cross sections for $^{30,32,34,35,37,38}Ba$ isotopes were calculated from 10 MeV to 20 MeV neutron energies using TALYS 1.6, EMPIRE-3.2.2, and ALICE. Ba, first isolated in 1808 by English chemist Sir Humphry Davy, is a chemical element with atomic number 56 and is a soft silvery metallic alkaline earth metal. There are seven naturally occurring isotopes of barium ($^{30,32,34,35,37,38}Ba$) [6]. The most common naturally occurring minerals of barium are $BaSO_4$ and $BaCO_3$, called as barite. Since radioactive isotopes of barium are among the comparatively abundant fission products, they have a biological importance [7]. The barium isotopes which is already found to be product of irradiated uranium are associated with one of the two nuclei formed by uranium fission [8].

Calculations were performed by using EMPIRE [9], TALYS [10], and ALICE [11] codes. These model codes are used for evaluating of many nuclear data in literature. The GDH model for pre-equilibrium emissions [12,13] and the W-E model for equilibrium calculations [14] are used in the ALICE.

In this study, semi empirical formulae developed by four different authors were also used for Ba isotopes at 14.5 MeV. These formulae have been in Table 1.

Table 1: Four different semi empirical formulae of the (n,2n) reaction at 14.5 MeV.

Author	Formula, $\sigma(n,2n)$ (mb)	Mass region
Tel et al. [15]	$=\exp\{7.15[1-2.45\exp(-31.62s)]\}$ $=\exp\{7.65[1-1.59\exp(-23.06s)]\}$	14≤A≤241 $\begin{cases} \text{for even } A \\ \text{for odd } A \end{cases}$
Goyal &Gur [16]	$=-69558.98(Z/A)^3 + 13430.42(Z/A)+1116.7$ $=25340(Z/A)^3-40114(Z/A)+16543$	$\begin{cases} \text{for even } A \\ \text{for odd } A \end{cases}$
Ikeda *et al.* [17]	$=\exp\begin{bmatrix}7.7590(1-0.80645 \\ \exp(-17.038(N-Z)/A))\end{bmatrix}$	For all A
Bychkov et al. [19]	$=(100+A)\begin{Bmatrix}1-\exp[-10.0(N-Z)/A] \\ [-5.7301+135.94(N-Z)/A]\end{Bmatrix}$ $=(100+A)\{3.3082+19.963(N-Z)/A\}$	$0.03 \le (N-Z)/A \le 0.11$ $(N-Z)/A \ge 0.11$

It is important to emphasize that a correction was made on the formula given by Tel et al. (2008) [15]. When the formula was multiplied by a factor of 1.4, the results were seen a good agreement with model calculations and experimental data at 14.5 MeV.

2 Results and Discussions

In this present work, $(n,2n)$ reaction cross sections were theoretically calculated for the reactions $^{130}Ba(n,2n)^{129}Ba$, $^{132}Ba(n,2n)^{131}Ba$, $^{134}Ba(n,2n)^{133}Ba$, $^{135}Ba(n,2n)^{134}Ba$, $^{136}Ba(n,2n)^{135}Ba$, $^{137}Ba(n,2n)^{136}Ba$, and $^{138}Ba(n,2n)^{137}Ba$ at an incident neutron energy of 14.2 MeV using their different models up to 20 MeV. These models are given as EMPIRE [8], TALYS [10], and

ALICE/ASH [11]. The evaluation data in ENDF/BVII.1 library and the experimental cross-section data from EXFOR [16] were also used for comparison.

The calculated cross-sections were shown in Figs. 1-7. Figs. 1-7 show theoretical EMPIRE 3.2 Malta, TALYS 1.6, ALICE/GDH calculations, and also show the evaluated ENDF/B-VII.1 data. The semi-empirical data and the ENDF/B-VII.1 data were used to compare with three calculated cross section code data. Moreover, the experimental data were also plotted in all figures around 14.5 MeV.

Theoretical calculations for the ^{130}Ba(n,2n)^{129}Ba reaction obtained from there computer codes using EMPIRE-3.2.2, TALYS-1.6, and ALICE/GDH in the energy range up to 20 MeV have been given in Fig. 1. Theoretical results have been compared with the evaluated nuclear data of ENDF/B-VII.1, the available experimental values from EXFOR [16], and semi-empirical calculation developed by Tel et al. [15], Goyal and Gur [16], Ikeda et al. [17], and Bychkov et al. [18]. From the Fig.1, all theoretical calculations and ENDF data near 14 MeV are in agreement with available measured data of Lu et al., Pepelnik et al. [16] and the data (1431.75mb) from Tel et al [15] while the experimental data from Konno et al. [16] were higher.

The theoretical cross section calculations of ^{132}Ba(n,2n)^{131}Ba reaction of EMPIRE-3.2.2, TALYS-1.6, and ALICE/GDH in the energy range from 10 MeV to 20 MeV were given in Fig. 2. The semi empirical formulae data by Tel et al. [15]. Sakane et al. [16] are also in approximately agreement with the results of ALICE/ASH and ENDF lines at energy range between 13 MeV and 15 MeV. However, Luo et al. and Kondo et al. have higher cross section values at the same energy range. The calculated cross section data from TALYS and EMPIRE are lower values.

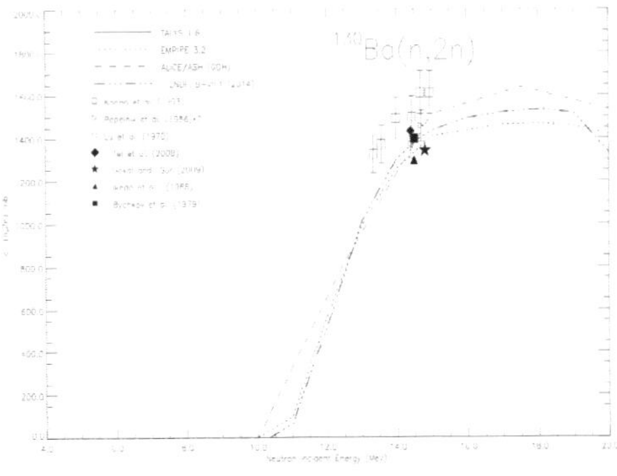

Fig.1. ^{130}Ba(n,2n)^{129}Ba reaction cross section.

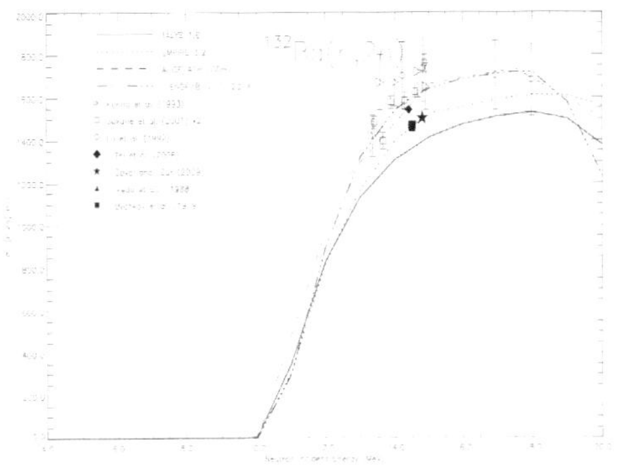

Fig.2. ^{132}Ba(n,2n)^{131}Ba reaction cross section.

The cross section results of ^{134}Ba (n,2n)^{133}Ba up to 20 MeV is shown in Fig. 3. As seen from Fig. 3, ALICE/GDH and EMPIRE lines show a close harmony with experimental data by Konno et al. and Xiangzhong et al. from EXFOR [16]. The agreement between calculated values from ENDF and TALYS-1.6 and results of the experimental data of Ikeda et al. from EXFOR [16] have similar values at 14-15 MeV. Moreover, the semi empirical calculated data using Table 1 are approximately same values with Ikeda et al. data and with ENDF data.

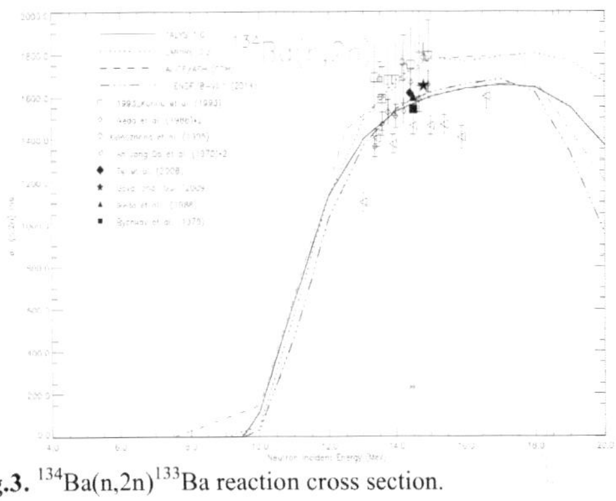

Fig.3. ^{134}Ba(n,2n)^{133}Ba reaction cross section.

The excitation function of the ^{135}Ba(n,2n)^{134}Ba up to 20 MeV was given in Fig. 4. From Fig. 4, the line from of ALICE/GDH seems to be higher than the other lines, especially between 6 and 18 MeV. The only one experimental data (Strohmaier et al.) [16] was found in literature. EMPIRE-3.2.2, TALYS-1.6, and ENDF/B-VII.1 data points are in agreement with each other and experimental data at all energy ranges. Strohmaier et al.'s measurements agree with the calculation between 7 to 20 MeV. All semi empirical calculations at 14.2 MeV agree with all calculated results except for ALICE-GDH data.

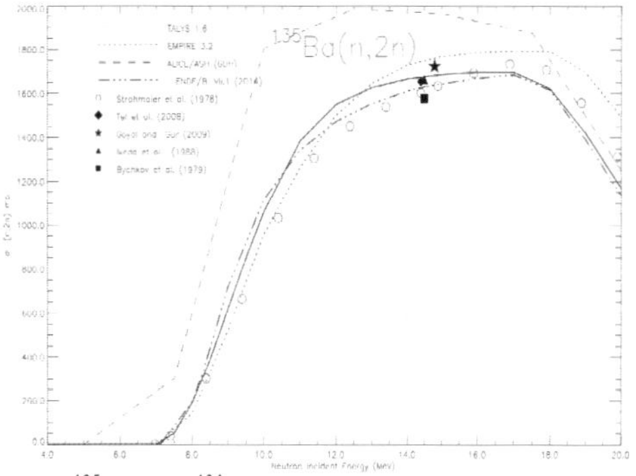

Fig.4. $^{135}Ba(n,2n)^{134}Ba$ reaction cross section.

Calculations of $^{136}Ba(n,2n)^{135}Ba$ up to 20 MeV was presented in Fig. 5. For $^{136}Ba(n,2n)$ reaction, three experimental data were found from literature. All experimental data were multiplied by a factor of 2. Ikeda et al.'s measurements agree with the calculation between 13.3 and 14.9 MeV. Strohmaier et al.'s measurements agree with the calculation below 11.5 MeV. Above 5 MeV, however, their data are much higher than the calculation. There is good agreement between the cross-section between 10 and 15 MeV calculated with three computer codes and ENDF/B-VII.1 files and the experimental data from EXFOR. Semi empirical data using formulae given in Table 1 at 14.2 MeV were also close agreement with both calculated data from EMPIRE, TALYS, and ENDF/B-VII.1 files and also experimental data.

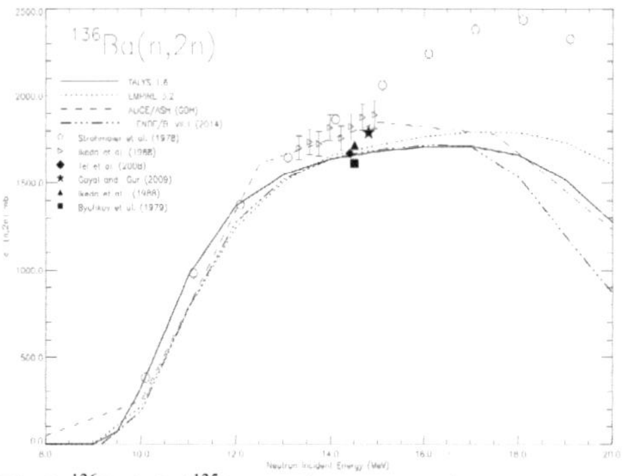

Fig.5. $^{136}Ba(n,2n)^{135}Ba$ reaction cross section.

The cross sections of $^{137}Ba(n,2n)^{135}Ba$ were given in Fig. 5. The theoretical calculations of EMPIRE-3.2.2, ALICE-GDH, and TALYS-1.6 codes, and ENDF/B-VII.1 give approximately good agreement with experimental and semi-empirical data except from Strohmaier et al. [16], the obtained semi empirical data from Tel et al. [15] was seen same value with theoretical data from EMPIRE at 14.5 MeV.

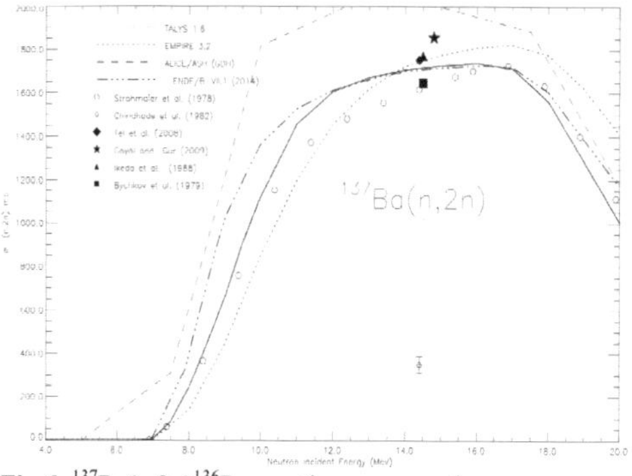

Fig.6. $^{137}Ba(n,2n)^{136}Ba$ reaction cross section.

The calculated $^{138}Ba(n,2n)^{137}Ba$ reaction cross sections and comparison with EXFOR database are shown in Fig. 7. As seen from Fig. 7, all theoretical lines from codes seem to be slightly same with experimental data and with ENDF/B-VII.1. The semi empirical data by Tel et al. [15] and Bychkov et al. are also close to data calculated with TALYS, while Ikeda et al. and Goyal and Gur are close to data calculated with ALICE-GDH.

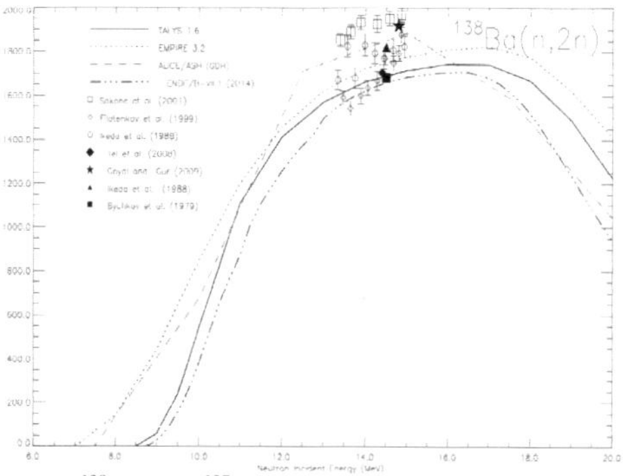

Fig.7. $^{138}Ba(n,2n)^{137}Ba$ reaction cross section.

3 Conclusion

In this work, we deduce excitation functions for the following reactions: $^{130}Ba(n,2n)^{129}Ba$, $^{132}Ba(n,2n)^{131}Ba$, $^{134}Ba(n,2n)^{133}Ba$, $^{135}Ba(n,2n)^{134}Ba$, $^{136}Ba(n,2n)^{135}Ba$, $^{137}Ba(n,2n)^{136}Ba$, and $^{138}Ba(n,2n)^{137}Ba$. Theoretical calculated results within TALYS, EMPIRE, and ALICE nuclear model codes were compared with evaluated nuclear data from EXFOR, ENDF data libraries and also compared with some published semi empirical formula based on the statistical model obtained from literature. The model calculations are seen in general in agreement with the experimental cross-sections from EXFOR and semi-empirical cross section data from literature.

References

1. E. Tel, S. Okuducu, A. Aydin, B. Sarer, G. Tanir, Acta Physica Slovaca, **54**, 191 (2004)

2. R.A. Forrest, J. Kopecky, Fusion Engineering and Design **82**, 73 (2007)
3. S.L. Goyal, P. Gur, Pramana-J. Phys. **72**, 355 (2009)
4. E. Tel, Journal Fusion Energy, **29**, 332 (2010)
5. H. Il Kim, M. Herman, S. F. Mughabghab, P. Oblozinsky, D. Rochman, Young-Ouk Lee, Nuclear Science and Engineering, **160**, 168 (2008)
6. K. Kurban, Z. Zhengjun, S. Xiuquan, H. Yinlu, S. Qingbiao, Communication of Nuclear Data Progress **25** (2001)
7. S.E. Agbemava, R.B.M. Sogbadji, Nyarko B.J.B., Della R., Annals of Nuclear Energy **38** (2011) 379–382
8. Iliffe C.E., An Introduction to Nuclear Reactor Theory, Published by Manchester University Press (1984)
9. M. Herman, et al., EMPIRE-3.2 Malta modular system for nuclear reaction calculations and nuclear data evaluation, User's Manual (2013)
10. A. Koning, S. Hilaire, S. Goriely, TALYS-1.6-A Nuclear Reaction Program, User Manual, 1st edition NRG, The Netherlands (2013)
11. C.H.M. Broeders, A. Yu. Konobeyev, Yu. A. Korovin, V. P. Lunev, M. Blann, Report FZKA 7183, http://bibliothek.fzk.de/zb/berichte/FZKA7183.pdf (2006)
12. M. Blann, H.K. Vonach, Phys. Rev. C, **28**, 1475 (1983)
13. Blann, M., Technical Report UCRL-JC-109052 (1991)
14. Weisskopf, V.F., Ewing, D.H., Physical Review 57, 472–485 (1940)
15. Tel, E. Okuducu Ş., Bölükdemir M. H., Tanır G., 2008. Semi-empirical systematics of (n,2n), (n,a) reaction cross sections at 14 15 MeV neutron energy, Int. J. Mod. Phys. E 17,1, 1-17.
16. http://www.nndc.bnl.gov/exfor/exfor.htm
17. Ikeda Y., Konno Ch., Oishi K., Nakamura T., Miyade H., Kawade K., Yamamoto H., Katoh T., JAERI 1312, Tokai-mura, (1988)
18. Bychkov, V.M., Manokhin, V.N., Pashchenko, A.B., Plyaskin, V.I., Voprosi Atomnoy Naukii Techniki. Jadernye Konstanty (Nucl.Constants) 1 (32), 27 (1979)

Monte Carlo calculation of proton stopping power and ranges in water for therapeutic energies

Ahmet Bozkurt

Akdeniz University, Faculty of Engineering, Department of Biomedical Engineering, Antalya, Turkey

Abstract. Monte Carlo is a statistical technique for obtaining numerical solutions to physical or mathematical problems that are analytically impractical, if not impossible, to solve. For charged particle transport problems, it presents many advantages over deterministic methods since such problems require a realistic description of the problem geometry, as well as detailed tracking of every source particle. Thus, MC can be considered as a powerful alternative to the well-known Bethe-Bloche equation where an equation with various corrections is used to obtain stopping power and ranges of electrons, positrons, protons, alphas, etc. This study presents how a stochastic method such as MC can be utilized to obtain certain quantities of practical importance related to charged particle transport. Sample simulation geometries were formed for water medium where disk shaped thin detectors were employed to compute average values of absorbed dose and flux at specific distances. For each detector cell, these quantities were utilized to evaluate the values of the range and the stopping power, as well as the shape of Bragg curve, for mono-energetic point source pencil beams of protons. The results were found to be ±2% compared to the data from the NIST compilation. It is safe to conclude that this approach can be extended to determine dosimetric quantities for other media, energies and charged particle types.

1 Introduction

While traversing through an absorbing material, a charged particle undergoes various interactions with the orbital electrons and the nuclei of the atoms in the medium. The amount of energy lost is usually characterized by the concept of stopping power which is defined as the rate of energy loss by a charged particle per unit path length in an absorbing medium. It is denoted as –dE/dx and usually has the units of MeV cm2/g when named as mass stopping power. This concept plays an important role in radiation dosimetry and protection when determining certain properties of charged particle beams such as the range. It depends on the type and energy of the charged particle, as well as the properties of the absorbing material it penetrates, such as its density and atomic number.

Before expending its kinetic energy, a charged particle experiences many different interactions that may change its path in the form of elastic or inelastic scatterings. The kinetic energy of the particle is also decreased in the process either through transfers to electrons (collision loss) or to bremsstrahlung photons (radiative loss). The total stopping power for charged particle interactions is composed of two parts; namely collision stopping power and radiative stopping power.

$$\left(-\frac{dE}{dx}\right)_{tot} = \left(-\frac{dE}{dx}\right)_{col} + \left(-\frac{dE}{dx}\right)_{rad} \quad (1)$$

The radiative stopping power is the result of Coulomb interaction between a charged particle and the nuclei of the absorber. Compared to collision losses with orbital electrons, this quantity is usually negligible, especially for heavy charged particles such protons, alpha particles, etc. Therefore, energy transfer from energetic charged particles to an absorber they traverse occurs mainly through Coulomb interactions of the charged particles with orbital electrons of the absorber atoms.

Theoretically, mass stopping power (in units of MeV cm2/g) of an absorber for a certain type of charged particle is given by the Bethe formula, developed in the realm of quantum mechanics and relativity. This formulation is usually refined for some density and shell corrections and is then read as

$$\left(-\frac{dE}{\rho dx}\right)_{col} = \frac{4\pi e^4}{m_e c^2}\frac{z^2}{\beta^2}\frac{ZN_A}{A}\left\{\ln\left(\frac{2m_e c^2\beta^2}{I(1-\beta^2)}\right) - \beta^2\right\} \quad (2)$$

where Z and z are the atomic number of the absorber and the incoming charged particle, respectively; A is the atomic mass of the absorber; N_A: Avagadro's number; e is the charge of an electron; $m_e c^2$ is the rest energy of an electron (0.511 MeV); β is the relative velocity of the incoming charged particle; I is the mean ionization potential of the absorber [1].

Bethe's equation is very useful for calculating stopping power and ranges of charged particles in any material if one is interested only in point energies. For practical uses, there are data tables (provided by NIST) or computer codes (such as SRIM) which base their calculations on this equation [2,3]. Fig. 1 shows a sample graph of mass collision stopping power of water against proton kinetic energy obtained from such data.

Corresponding author: abozkurt@akdeniz.edu.tr

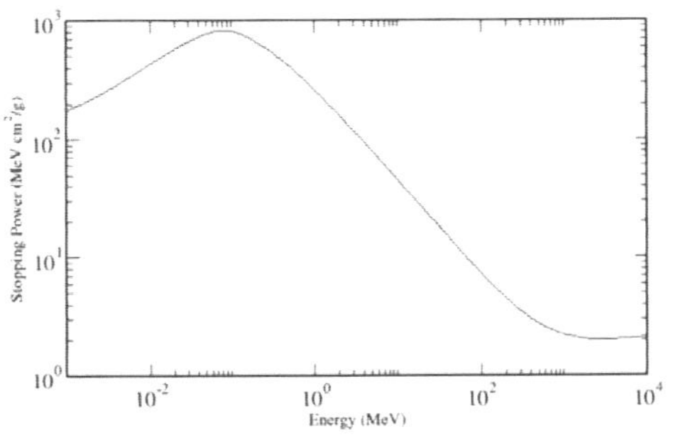

Fig. 1. Mass collision stopping power of liquid water for protons (adopted from NIST).

The cumbersome nature of Bethe's equation can be avoided by use of other techniques. This study presents an alternative stochastic method to compute the stopping power and ranges of protons in water using the well-known Monte Carlo simulations for proton energies encountered in therapeutic settings. A comparison with NIST data is also provided.

2 Materials and Methods

Monte Carlo is a well-established statistical technique for obtaining numerical solutions to physical or mathematical problems when physical measurements are either difficult or impossible. In charged particle transport problems, MC uses random numbers along with probability distributions to estimate the energy, position, direction and path-length of individual particles, as well as the type of physical interactions that particles experience when traversing a medium. The technique can yield quantities such as particle fluence across a surface or a cell and energy deposition in defined volumes. Thus, MC presents many advantages over deterministic methods since such problems require a realistic description of the problem geometry, as well as detailed tracking of every source particle. Thus, MC can be considered as a powerful alternative to the well-known Bethe-Bloche equation for modeling charged particle interactions [4].

Various Monte Carlo packages are available today that are specifically developed for handling radiation transport problems. MCNPX (version 2.7.0) was used in this study to model the geometry and composition of a water cylinder. [5] This Monte Carlo code was developed at the Los Alamos National Laboratory (LANL, USA) and is available from the Oak Ridge National Laboratory (ORNL). Transport and interaction of various particles can be handled in a wide range of energies and three-dimensional arbitrary geometries. Particle interactions are treated either as continuous or discrete energies where the cross-section data are adopted from the ENDF/B libraries (Evaluated Nuclear Data Files, version B). MCNPX can facilitate a wide variety of source distributions, detector conditions and user-configurable output options, as well as providing a rich set of variance reduction techniques [6].

A water cylinder (r=30 cm; h=100 cm) was modeled in a vacuum sphere (r=150 cm). Starting from the surface of this water phantom, 1000 concentric detector disks (r=2 cm) were created to record particle data. The thicknesses of these disks were determined approximately from a fraction of proton range at the source energy.

In the MCNP model, the mono-energetic protons were emitted mono-directionally from a point source at one end of the water cylinder. Tally types F4 and +F6 were placed in every detector cell in the region of interest to record the average cell flux (in units of #/cm^2/source particle) and the absorbed dose (in units of MeV/g/source particle) by all primary and secondary particles, respectively.

All the secondary particles that may arise from proton interactions were transported in the runs. The simulations were carried out for 1 billion proton histories that yielded better than 1% statistical errors. Overall, the precision in the results was satisfactory considering the extremely time-consuming nature of the problem. The typical CPU time was about 1000 min.

3 Results

MCNPX was used to compute absorbed dose and flux for protons in the energy range of 50-250 MeV at varying depths of water. For each incoming proton energy, the first detector was taken as the basis for stopping power calculations where the absorbed dose from +F6 tally (in units of $MeV/g/particle$) was divided by the flux from F4 tally (in units of $1/cm^2/particle$) to represent the stopping power ($MeV\ cm^2\ /g/particle$) of these protons in water. Later, the listing of the detector cells was searched to find the depth at which the absorbed dose falls below 1% of the maximum. This depth value was taken to represent the range of the incoming proton energy. Finally, the Bragg curve was sketched for each proton energy as a graph of relative dose (%) versus absorber depth ($cm^2\ /g$).

The calculated data of water obtained this way are given in Table 1 both for stopping power and range of 50-250 MeV protons. As expected, the stopping power of water for protons exhibits a decreasing relationship with increasing particle energy. This correlation can be approximated by a power equation of the form ($R^2 =$ 0.997)

$$S_p\left(\frac{MeV\ cm^2}{g}\right) = 206.7[E(MeV)]^{-0.72}. \quad (3)$$

On the other hand, the range of protons in water displays a direct proportionality with particle energy which can be summarized by this fit equation ($R^2 =$ 0.999)

$$R(g/cm^2) = 0.0023[E(MeV)]^{1.76} \quad (4)$$

The results obtained in this study using MCNPX simulations were compared with NIST values and found good agreement within %4 for stopping power and %0.2 range data. Fig. 2 is a plot of simulation results and NIST tables as a function of proton energy.

DOI: 10.1051/epjconf/201715401007

Table 1. Stopping power and range of protons in water.

E (MeV)	Stopping power $(MeV\ cm^2/g)$ MCNPX	Stopping power $(MeV\ cm^2/g)$ NIST	Difference (%)	Range (g/cm^2) MCNPX	Range (g/cm^2) NIST	Difference (%)
50	12.72	12.45	-2.2%	2.231	2.227	-0.2%
60	11.13	10.78	-3.2%	3.096	3.093	-0.1%
70	9.751	9.559	-2.0%	4.088	4.080	-0.2%
80	8.502	8.625	1.4%	5.195	5.184	-0.2%
90	7.989	7.888	-1.3%	6.408	6.398	-0.2%
100	7.516	7.289	-3.1%	7.731	7.718	-0.2%
125	6.282	6.192	-1.5%	11.48	11.46	-0.2%
150	5.551	5.445	-1.9%	15.80	15.77	-0.2%
175	4.923	4.903	-0.4%	20.65	20.62	-0.1%
200	4.662	4.492	-3.8%	25.99	25.96	-0.1%
225	4.210	4.170	-1.0%	31.80	31.74	-0.2%
250	4.001	3.911	-2.3%	38.01	37.94	-0.2%

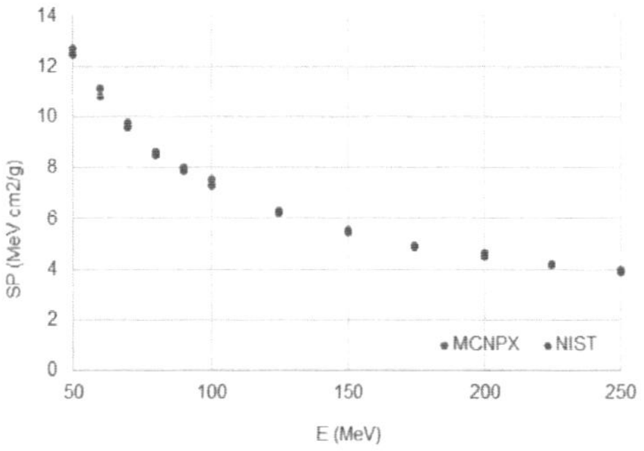

(a) Stopping power

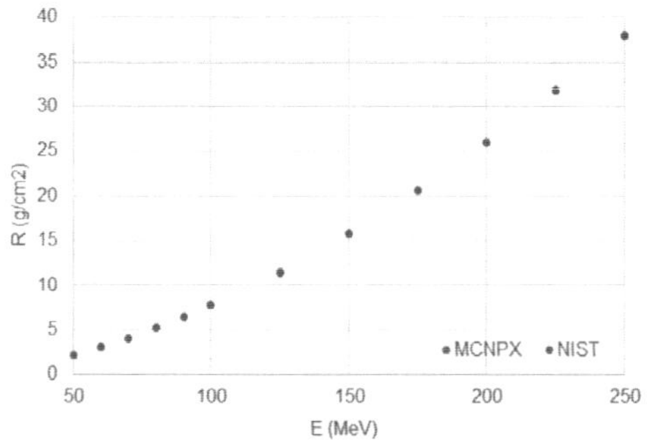

(b) Range

Fig. 2. Stopping power and range of protons in water as a function of incoming proton energy

4 Conclusion

This study aims to conceptualize a Monte Carlo approach for calculating stopping power and range of charged particles. The results obtained from MCNPX simulations of 50-250 MeV protons in water medium suggest that stopping power of a mono-energetic proton beam can be calculated from the computed values of absorbed dose and cell flux. This approach is in good agreement with the tabular values provided in NIST databases based on computer codes that utilize Bethe's equation. The results are encouraging in that detailed studies that involve wider energy range, different particle types, and various materials need to be carried out to investigate the validity of this methodology.

References

1. F.H. Attix. *Introduction to Radiological Physics and Radiation Dosimetry*. John Wiley & Sons (1986)
2. NIST, www.physics.nist.gov/PhysRefData/Star/Text/contents.html, Last accessed: 01.05.2017
3. SRIM – The stopping and range of ions in matter., Nuclear Instruments and Methods in Physics Research Section B: Beam Interactions with Materials and Atoms, **268**(11-12):1818-1823 (2010)
4. P. Andreo. Monte Carlo techniques in medical radiation physics. Phys. Med. Biol.; **36**(7):861-920 (1991).
5. J.F. Briesmeister MCNP-A general Monte Carlo N-particle transport code, Version 4B. Los Alamos National Laboratory Report; LA-12625-M (1997).
6. L.S. Waters. MCNPX User's Manual, Version 2.3.0, LA-UR-02-2607 (2002).

Corresponding author: abozkurt@akdeniz.edu.tr

Comparison of 3DCRT,VMAT and IMRT techniques in metastatic vertebra radiotherapy: A phantom Study

Sonay GEDIK, *Sema* TUNC, *Arda* KAHRAMAN, *Sibel* KAHRAMAN CETINTAS and *Meral* KURT

Uludag University, Medicine Faculty, Radiation Oncology Department, Bursa, Turkey

Abstract. Vertebra metastases can be seen during the prognosis of cancer patients. Treatment ways of the metastasis are radiotherapy, chemotherapy and surgery. Three-dimensional conformal therapy (3D-CRT) is widely used in the treatment of vertebra metastases. Also, Intensity Modulated Radiotherapy (IMRT) and Volumetric Arc Therapy (VMAT) are used too. The aim of this study is to examine the advantages and disadvantages of the different radiotherapy techniques. In the aspect of this goal, it is studied with a randophantom in Uludag University Medicine Faculty, Radiation Oncology Department. By using a computerized tomography image of the phantom, one 3DCRT plan, two VMAT and three IMRT plans for servical vertebra and three different 3DCRT plans, two VMAT and two IMRT plans for lomber vertebra are calculated. To calculate 3DCRT plans, CMS XiO Treatment System is used and to calculate VMAT and IMRT plans Monaco Treatment Planning System is used in the department. The study concludes with the dosimetric comparison of the treatment plans in the spect of critical organ doses, homogeneity and conformity index. As a result of this study, all critical organ doses are suitable for QUANTEC Dose Limit Report and critical organ doses depend on the techniques which used in radiotherapy. According to homogeneity and conformity indices, VMAT and IMRT plans are better than one in 3DCRT plans in servical and lomber vertebra radiotherapy plans.

1 Introduction

Treatment rate is increasing for elderly patients at 65 and more ages with the increasing population age today. It is expected that 70 percent of all cancer patients loses their lives in this age range [1]. It is estimated that this death rate will be doubled by 2030 [2,3]. Until 2050, 65 years of age and over are expected to receive four times more new cancer diagnoses [2,3]. Vertebral metastasis is an important morbidity source [4]. Vertebra is the most frequent site of body after lung and liver [5]. Approximately one-third of cancer patients develop bone metastasis, and 10 percent of these metastases are vertebra metastases [6]. For example, 10 percent of colorectal cancers have bone metastasis [7]. According to a study by Walther of 1948 on the autopsy of 3500 patients who lost their lives due to cancer, sixty six percent of all prostate cancers, fourty percent of breast cancer patients and thirty eight percent of thyroid cancer patients bone metastasis developed [8]. Patients with vertebral metastasis are palliative and treatment purpose is to control pain, to preserve or treat neurological functions, to prevent bladder involvement and to keep the patient standing [4]. The treatment of patients with vertebral metastases is decided by considering several factors. These factors are; clinical data, duration of present symptoms, tumour type, radiosensitivity, tumour location, non-vertebral extension, vertebra length and medical status of the patient [4]. Techniques such as Intensity Modulated Radiotherapy (IMRT) and Volumetric Arc Therapy (VMAT) are also used in the radiotherapy of vertebral metastases, as well as conventional planning techniques. In 3-Dimensional Conformal Radiotherapy (3DCRT), the doses of healthy tissues and organs are high. The IMRT and VMAT methods allow the dose to accumulate in tumour volume, preserving normal tissues. This feature is very important for the patient in terms of the coming time. Nonetheless, it can lead to radiation-induced secondary cancers. The aim of this study is to answer the question of which treatment planning method is best suited to the application of radiotherapy and the most appropriate for critical organs in patients with vertebral metastases from 3DCRT, IMRT and VMAT treatment methods.

1.1 Treatment approaches to metastatic vertebras

Bone metastases may be treated with external radiotherapy, radionuclides, surgery, chemotherapy and co-administration of one or more of the bisphosphonates [9].

Vertebra metastases are a common occurrence in the cancer prognosis. The methods of treatment are both radiotherapy or surgery or both of them. The use of 3DCRT in the radiotherapy of vertebra metastases is quite common. However, IMRT and VMAT methods can also be used.

3DCRT is a technique performed by matching radiation beams to tumour shape. Conventional external radiotherapy treatment is performed with the aid of a compensator or wedge to change the intensity or density of the radiation bundles at the single density throughout the treatment area. The treatment modality by changing the bundle density and obtaining a composite plan is called IMRT. VMAT is the application of IMRT (step and shoot and dynamic methods) where the treatment gantry unit is rotating with arcs. Treatment can be done with more than one arc or partial arc. In the literature, it is applied as 30 Gy with 10 fractions, 20 Gy with 5 fractions and 8 Gy with 1 fractions in world-wide metastatic vertebra radiotherapy.

1.2 Materials and Reference Values

1.2.1 Randophantom

The randophantom produced from tissue equivalent material consists of sections parallel to each other with a thickness of 2.5 cm. These sections are connected to each other by removable inner support rods. Its section has pits closed to the bone equivalent to soft tissues or lung tissue. These holes are the designed so that the pits are removed and replaced with a dosimeter. Rando phantom represents a woman weighing 55 kg, 155 cm in length.

1.2.2 Conformity and Homogeneity Indexes

In the quality assessment of treatment plans, the homogeneity and conformity index formulas specified in the 83rd Report of ICRU (International Commission on Radiation Units and Measurements) were used [10].

$$HI = \frac{D2\%-D98\%}{D50\%} \quad (1)$$

D2% : Doses of 2% of PTV volume
D98%: Doses of 98% of PTV volume
D50%: Doses of 50% of PTV volume

$$CI = \frac{TVRI}{TV} x \frac{TVRI}{VRI} \quad (2)$$

***TV_{RI}*:** Target volume receiving reference isodose
***TV*:** Target volume
***V_{RI}*:** Total volume receiving reference isodose of the target volume and non-target volume

According to ICRU Report No. 83, if the homogeneity index value is close to 0 and the conformity index value is close to 1, it is desirable to calculate the optimum treatment plan.

1.2.3 Dose Limitations

The dose-volume limits of QUANTEC (Quantitative Analysis of Normal Tissue Effects in the Clinic) were taken into account when determining critical organ doses.

2 Techniques

This study was carried out using randophantom belonging to Uludag University Medical Faculty, Radiation Oncology Department. Rando phantom was computerized tomography and computed with cervical and lumbar vertebral 3D-CRT, VMAT and IMRT treatment planning systems on this image. The total number of plans for both regions is 13. The results obtained with these treatment plans are shown on the charts with the dose of the risky organs and the dose given to the target volunteer with the help of DVH.

The randophanton CT image was taken at a 3 mm slice thickness on the supine position in the head first position.

PTVs are created by 0.3 cm margins for CTV Cervical and CTV Lumber volumes. All VMAT plans were made with 6MV photons. In 3DCRT plans, both 6 MV and 15 MV photons were used. It is applied as 30 Gy with 10 fractions for both servical and lumber vertebra in this study.

These treatment plans were calculated and abbreviations used to classify the data obtained by the calculated treatment plans are as follows:

S-3DCRT: Conformal radiotherapy treatment plan for cervical vertebra, calculated by using two treatment fields at right and left laterals
S-VMAT1: Radiotherapy treatment plan calculated by using two halves and one full arc beams for the cervical vertebra
S-VMAT2: Radiotherapy treatment plan for a cervical vertebra, calculated by using a full arc beam
S-IMRT1: IMRT treatment plan with seven treatment beams at 0, 52, 104, 156, 208, 260, 302 degrees for cervical vertebra
S-IMRT2: IMRT treatment plan with seven treatment beams at 110, 130, 150, 180, 210, 230 and 250 degrees for the cervical vertebra
S-IMRT3: IMRT treatment plan with seven treatment beams at 110, 130, 150, 180, 210, 230 and 250 degrees for cervical vertebrae, 15 degree angle of collimator angle in 110 degree and 250 degree treatment areas.
L-3DCRT1: Conformal radiotherapy treatment plan calculated for the Lomber vertebra using AP and PA fields
L-3DCRT2: Conformal radiotherapy treatment plan for the Lomber vertebra, calculated by intervening AP and PA fields

DOI: 10.1051/epjconf/201715401008

L-3DCRT3: Conformal radiotherapy treatment plan for Lomber vertebra, calculated by placing one AP and two posterior oblique treatment areas
L-VMAT1: Radiotherapy treatment plan calculated by using two half and one full arc beams for the lomber vertebra
L-VMAT2: Radiotherapy treatment plan for the lomber vertebra, calculated by using a full arc beam
L-IMRT1: IMRT treatment plan with seven treatment beams at 0, 52, 104, 156, 208, 260, 302 degrees for the lomber vertebra
L-IMRT2: IMRT treatment plan with seven treatment beams at 110, 130, 150, 180, 210, 230 and 250 degrees for the lomber vertebra

3 Results

OAR doses calculated in treatment plans are shown in Table1 and Table 2.

Table 1. Risky organ doses in 3D-CRT and VMAT treatment plans for servical vertebra.

		S-3DCRT	S-VMAT1	S-VMAT2
Thyroid	V_{26}	13.9	22.79	25.83
Larynx	D_{mean}(cGy)	2743	2791	2811
OralCavity	D_{mean}(cGy)	112	1244	1350
Mandible	D_{max} (cGy)	3157	2899	2835
SpinalCord	D_{max} (cGy)	3092	3186	3206

Table 2. Risky organ doses in IMRT treatment plans for servical vertebra.

		S-IMRT1	S-IMRT2	S-IMRT3
Thyroid	V_{26}	23.67	11.68	11.21
Larynx	D_{mean}(cGy)	2753	2819	2820
OralCavity	D_{mean}(cGy)	1138	1477	1465
Mandible	D_{max} (cGy)	2939	2505	2575
SpinalCord	D_{max} (cGy)	3229	3225	3210

Table 3. Risky organ doses in 3D-CRT treatment plans for lumber vertebra.

		L-3DCRT1	L-3DCRT2	L-3DCRT3
Kidney(L)	V_{12}	22.18	19.69	28.5
	V_{18}	15.21	12.03	17.08
	V_{20}	13	9.61	14.14
Kidney(R)	V_{12}	24.61	19.47	29.67
	V_{18}	16.92	11.74	12.38
	V_{20}	14.5	9.38	8.51
Liver	V_{20}	13.14	9.61	9
SpinalCord	D_{max}(cGy)	3212	3199	3339

Table 4. Risky organ doses in VMAT and IMRT treatment plans for lumber vertebra.

		L-VMAT1	L-VMAT2	L-IMRT1	L-IMRT2
Kidney (L)	V_{12}	45.2	37.61	44.37	45.87
	V_{18}	17.9	15.99	19.92	22.53
	V_{20}	13.35	12.60	14.64	17.43
Kidney (R)	V_{12}	33.97	36.68	23.46	27.65
	V_{18}	16.15	16.53	10.24	13.4
	V_{20}	12.07	12.44	7.27	9.94
Liver	V_{20}	0.7	0.57	0.66	0.33
Spinal Cord	D_{max}(cGy)	3220	3243	3285	3282

4 Conclusions

Generally, the homogeneity index value is found to be the best in treatment planning calculated by VMAT and IMRT techniques.

The optimal conformity index value in cervical vertebra treatment planning was obtained by treatment plans calculated by VMAT and IMRT techniques. All plans are consistent with this limitation given the dose limits for the thyroid described in QUANTEC. But the most appropriate treatment plan is the S-IMRT3 treatment plan, according to the V26 dose. The S-3DCRT and S-IMRT2 treatment plans have a close thyroid dose with the S-IMRT3 treatment plan. It is thought that the best result of thyroid V26 in IMRT treatment plans is caused by gantry and collimator angles where the treatment fields used in S-IMRT2 and S-IMRT3 treatment plans are formed. In the S-3DCRT treatment plan, the treatment areas were created from the right and left laterals, allowing for thyroid protection.

In cervical vertebra radiotherapy, spinal cord dose is very important. According to the dose limits stated in QUANTEC, the maximum dose of spinal cord should be below 45 Gy. If the dose prescribed in all plans is below 45 Gy, the most appropriate treatment plan for this dose limitation was obtained with the S-3DCRT treatment plan. While the S-3DCRT treatment plan followed the maximum dose S-VMAT1 treatment plan received by the spinal cord, there were no significant differences between the other VMAT and IMRT treatment plans.

Significant differences were observed between treatment plans in terms of MU per fraction. It has been observed that the MU value per fraction seen in treatment plans calculated with the 3DCRT technique is about half of the value seen in treatment plans calculated by VMAT and IMRT technique. It has been observed that VMAT treatment plans are higher in terms of MU value per fraction seen in treatment plans calculated with the IMRT technique.

The best homogeneity index value was obtained from the lumbar vertebra treatment plan calculated by VMAT technique. The treatment plans calculated by the VMAT technique are followed by the treatment plans calculated by the IMRT technique. The highest homogeneity index value was calculated in the L-3DCRT3 treatment plan.

The homogeneity index value in treatment planning, which is calculated for the lumbar vertebra, is ideal for treatment planning computed by the VMAT and IMRT technique.

Conformity index value is above 1 value for all treatment plans calculated for the lumber vertebra. Conformity index values are obtained from treatment plans obtained with the best IMRT technique in order to be close to 1 value.

References

1. R.Yancik, I. Ries, Hematological Oncology Clinic North America,14:17-23 (2000).
2. J.A. Hayman, K.M. Langa, M.U. Kabeto, J ClinicOncol 19:3219;-3225 (2001)
3. A. Jatoi, S. Hillman, P. Stella, J Clin Oncol 23:9113-9119 (2005)
4. D.R. Fourney, Z.L. Gökaslan, Neurosurgery Clinic doi:10.1012/j.nec.2004.04.008 (2004)
5. D. Lubgan, A. Ziegaus, DOI:10.1007/s00066-014-0780-4
6. D.A. Wong, V.L. Fornasier, I. MacNab 15:1-4 (1990)
7. D. Santini, Ann Oncol ;23:2072-7 (2012)
8. H.E. Walther, Bens Schwabe Verlag (1948)
9. Kemik Metastazları,Temel ve Klinik Radyoterapi Kitabı,*Türk Radyasyon Onkolojisi Derneği*,s.387.
10. ICRU Report 83: Prescribing ,Recording,adn Recording Photon-Beam Intensity Modulated Arc Therapy,Vol.10,No:1 (2010)

DOI: 10.1051/epjconf/201715401009

Analysis of natural radioactivity in Yatağan coal – fired power plant in Turkey

Aydan Altıkulaç[1], *Şeref* Turhan[2] and *Hasan* Gümüş[3]

[1]Muğla Sıtkı Koçman University, Ula Ali Koçman MYO, Department of Electricity and Energy, Muğla, Turkey
[2]Kastamonu University, Faculty of Sciences and Art, Department of Physics, Kastamonu, Turkey
[3]Ondokuzmayıs University, Faculty of Sciences and Art, Department of Physics, Samsun, Turkey

Abstract. Use of the coal in order to generate electricity increases the exposure of people to radiation. In this paper, the activity concentrations of nuclides ^{226}Ra, ^{232}Th and ^{40}K in samples of coal and bottom ash from the Yatagan Coal–Fired thermal power plant determined using gamma ray spectrometer with a NaI(Tl) scintillation detector. The mean activity concentrations of ^{226}Ra, ^{232}Th, and ^{40}K in the coal were found to be 37.2±2.8 Bqkg^{-1}, 51.8±3.4 Bqkg^{-1} and 166.7±11.1 Bqkg^{-1}, respectively. Whereas in the bottom ashes, the concentrations of the corresponding radionuclides were found to be 62.2±5.6 Bqkg^{-1}, 87.4±5.9 Bqkg^{-1} and 221.0 ±12.5 Bqkg^{-1}, respectively. The findings show that bottom ashes show higher activity concentrations of related radionuclide to coal samples. The absorbed gamma dose rate in outdoor air DR_{OUT} and annual effective dose rate (AED) from coal were calculated to define radiological risk. The average findings of annual effective doses were detected as 68.6±5.1 μSvy^{-1} and 110.3±11.2 μSvy^{-1}, respectively.

1 Introduction

Energy is one of the most important inputs to the economic and social development of a country. Much of the energy needed in the world is met by fossil sources such as coal and natural gas. The most important share of fossil energy resources in Turkey belongs to the coal mine. According to World Energy Council statistics, world coal reserves freeze to 861 billion ton [1]. About 1.7% of the world's coal reserves are in Turkey. Coal is used as fuel in coal operated power plants to generated electricity in Turkey. Coal and lignite fueled thermal power plants in Turkey reported total electricity generation of 90.335 GWh per year. In the composition of fuels such as coal, there are fewer or more natural radioactive substances, such as ^{226}Ra, ^{210}Po and ^{210}Pb, as well as ^{40}K, derived from ^{238}U and ^{232}Th. The uranium and other radioisotopes in the lignite coal burned at high temperatures in thermal power plants becomes enrich and passes to the ash. Part of the heavy ash falls down to the bottom of the furnace as "bottom ash" together with unburned organic matter, while light ashes are transported along with hot gases to the pail and form fly ash. The average activity concentrations of ^{226}Ra, ^{232}Th and ^{40}K in coal according to UNSCEAR are 20, 20 and 50 Bqkg^{-1}, respectively [2-4]. The fly ash formed by burning the coal causes the radioactive materials to spread in the atmosphere. A large proportion of the radioactivity is remains in the bottom ash after combustion. About 0.5-2 % of the total ash resulting from the fly ash combustion is released to the atmosphere and this event creates a potential risk for the very close environment of the coal fired power plant [5]. The amount of coal required to produce 1 GWh of electricity per year is 3×10^9 kg. As a result, the radioactive pollution given to the atmosphere per year is 4000 MBq for ^{40}K, 1500 MBq for ^{238}U through ^{226}Ra, 5000 MBq for ^{210}Pb and ^{210}Po and 1500 MBq for the radionuclides of ^{232}Th decay series, from ^{232}Th to ^{224}Ra. As well as there are the ash mountains that generate atmospheres ^{222}Rn around each coal –fired power plant. Especially radioactive radon (^{222}Rn) gas and short-lived decay products (^{214}Bi of ^{214}Pb etc.) can be taken into the body through the respiratory tract and increase the risk of cancer by damaging the lung tissue [6]. In this study, the activities of natural radioactive elements were determined in the samples of coal and bottom ash.

2 Material and method

2.1 Preparing samples for measurement

Yatağan Thermal Power Plant has 3x210 MW power is located in Muğla, which has a significant portion of Turkey's lignite reserves [7]. In this paper, radioactivity measurements were made in the samples of bottom ash generated by burning of coal and coal used as fuel in Yatağan Thermal Power Plant. All samples were dried for one day in the oven at a preheated of 100 degrees. The dried samples were put in standard vessels and the mass and date of creation were recorded. Each sample vessel was wrapped with parafilm to provide a seal.^{226}Ra and ^{232}Th in all samples were expected to reach a radioactive equilibrium for a month.

2.2 Gamma measurements

Gamma spectrometric measurements of ^{226}Ra, ^{232}Th and ^{40}K in the samples were carried out by the agency of scintillation gamma-ray spectrometer with a sodium iodide (NaI) detector. The calibration of the detector performed using photo peaks from radioactive standards IAEA- RGU-1, IAEA-RGTh-1 and IAEA-RGK-1 reference materials. The detector has an energy resolution of 46 keV at 622 keV ^{137}Cs. The ^{226}Ra activity of the samples was detected by means of the density of 351.9 keV and 609.3 keV gamma lines of ^{214}Pb and ^{214}Bi, respectively. ^{232}Th activity was acquired through the gamma publisher of ^{228}Ac at 911.21 keV. ^{40}K concentration was directly read from spectrum analysis results. Spectrum analyzes were carried out using the ScintiVision gamma ray software.

3. Result and discussion

3.1 Activity concentration

The activity concentrations of ^{226}Ra, ^{232}Th and ^{40}K were calculated for 20 different coal and 20 different bottom ash sample. The results for coal samples are shown in Table 1. As seen in Table 1, the activity concentrations of ^{226}Ra,^{232}Th and ^{40}K were measured as 37.2±2.8 Bqkg^{-1}, (range from 26 to 47), 51.8±3.4 Bqkg^{-1}, (range from 19 to 80) and 166.7±11.1 Bqkg^{-1} (range from 137 to 189), respectively. The measured radioactivity and related radiological parameters in the bottom ash are as in Table 2. So, it can be said that bottom ashes show higher activity concentrations of related radionuclide to coal samples. The data obtained is higher than the data in UNSCEAR report [2,3].

3.2 Equivalent radium activity

This parameter is calculated by multiplying the corresponding radioactive elements by different coefficients and this calculation is shown in the following Equation [8,9].

$$Ra_{eq} = A_{Ra} + \frac{10}{7} A_{Th} + \frac{10}{130} \quad (1)$$

Where A_{Ra}, A_{Th} and A_K are the activities of ^{226}Ra, ^{232}Th and ^{40} K , respectively. As seen in Table 1 and Table 2, the equivalent activities calculated for bottom ash are about twice that of coal. Where A_{Ra}, A_{Th} and A_K are the activities of ^{226}Ra, ^{232}Th and ^{40} K, respectively.

As seen in Table 1 and Table 2, the equivalent activities calculated for bottom ash are about twice that of coal.

Table 1. Activity concentrations and calculated gamma doses in coal samples

Sample ID	^{226}Ra (Bqkg-^{1})	^{232}Th Bqkg-^{1})	^{40}K (Bqkg-^{1})	Ra_{eq} (Bqkg-^{1})	External absorbed gamma dose D_R (nGyh^{-1})
C2	29±2	19±1	178±11	68 ±6	32±2
C3	49±4	40±3	168±10	118±9	54±4
C4	37±2	23±1	186±12	83±6	39±3
C5	36±2	48±3	162±10	115±8	53±4
C6	40±3	52±3	189±13	128±9	58±4
C7	45±4	57±4	169±11	138±10	63±5
C8	39±3	53±3	137±10	124±9	56±4
C9	36±3	46±3	144±10	111±9	51±4
C10	47±3	60±4	180±12	145±11	66±5
C11	32±2	45±3	168±11	108±8	49±3
C12	36±3	48±3	156±11	115±8	53±4
C13	38±2	51±3	173±11	125±9	56±4
C14	40±3	54±3	188±12	130±10	59±4
C15	35±2	69±4	154±10	144±12	65±5
C16	39±2	77±5	172±11	160±13	73±5
C17	40±3	75±4	171±11	158±12	72±5
C18	40±4	80±4	137±9	163±12	74±6
C19	26±2	49±3	180±12	97±8	50±3
C20	33±2	54±3	153±10	120±9	51.±4
Average	**37±2**	**51±3**	**166±11**	**121±9**	**56±4**

3.3 Calculation of gamma doses

The gamma dose taken in the air (D_R) at a height of 1 m from the ^{226}Ra, ^{232}Th, and ^{40}K found in the structure of the material was calculated by the following Equation [10].

$$D(nGyh^{-1})=0.462A_{Ra} + 0.623A_{Th} + 0.0414A_K \quad (2)$$

where A_{Ra}, A_{Th} and A_K are the activity concentrations of ^{226}Ra, ^{232}Th, and ^{40}K in Bqkg^{-1}, respectively. External absorbed gamma dose rate calculated for coal and bottom ash samples are shown in Table 1 and Table 2. The calculated average value for coal is 56.9±4.4 nGyh^{-1} while the calculated value for bottom ash is 90.8±6.5 nGyh^{-1}. The dose rate in airdoors from terrestrial gamma rays in normally is around 60 nGyh^{-1}, with the mean range from 10 to 200 nGyh^{-1}, according to the UNSCEAR 2000 report [11].

The calculated results for coal and bottom ash samples are 68.6±5.1 and 110.3±11.2 μSvy^{-1}, respectively. According to the UNSCEAR 2000 report, this value was announced as 70 μSvy^{-1} for the world average [11]. It can be said that the mean value calculated for coal is around the world average. But the mean value for bottom ash is above the world average.

Table 2. Activity concentrations and calculated gamma doses in bottom ash samples.

Sample ID	^{226}Ra (Bqkg-^{1})	^{232}Th (Bqkg-^{1})	^{40}K (Bqkg-^{1})	Ra$_{eq}$ (Bqkg-^{1})	External absorbed gamma dose D$_R$ (nGyh^{-1})
BA1	47±3	34±2	241±14	112±6	53±3
BA2	48±3	50±3	239±14	136±8	62±4
BA3	60±5	49±3	250±14	148±8	74±4
BA4	64±6	76±5	267±15	191±11	87±5
BA5	74±6	99±6	220±12	230±12	103±7
BA6	65±5	84±6	204±12	200±11	89±6
BA7	63±5	86±6	172±9	197±11	88±6
BA8	68±5	92±6	203±11	213±12	95±6
BA9	60±5	80±5	176±10	187±10	83±5
BA10	59±4	77±5	191±11	183±10	81±5
BA11	59±4	79±5	202±11	186±10	84±6
BA12	72±6	98±6	194±10	226±13	100±7
BA13	70±6	93±6	227±13	219±12	98±7
BA14	59±4	76±5	253±14	185±11	83±5
BA15	71±6	132±8	237±13	276±13	122±8
BA16	58±3	112±7	256±14	249±14	105±7
BA17	50±3	96±6	219±12	202±12	90±6
BA18	55±3	115±7	234±13	235±13	104±7
BA19	62±5	104±6	216±12	225±12	100±7
BA20	74±6	117±7	227±12	256±14	114±8
Average	62±5	87±5	221±12	200±11	90±6

4 Conclusions

Muğla is located in Turkey, where has a significant portion of lignite reserves, and, ^{226}Ra, ^{232}Th, and ^{40}K radioisotopes are naturally found in these lignite coals by various measurements. The purpose of this study is to make measurements of natural radioactivity in samples of coal and bottom as samples taken from Yatağan Coal-Fired Power Plant basin in Muğla. The results show that the natural radioactivity of the bottom ash is about 2 times that of the coal natural radioactivity and this is an expected result. It can be said that the concentrations of radium, thorium and potassium in coal and bottom ash samples are slightly above the world average in the finds obtained.

References

1. World Energy Concil, Survey of Energy Resources, London. (2010)
2. UNSCEAR Ionizing Radiaiton: Sources and biological effects. United Nations Scientific Commitee on the Effects of Atomic Radiaiton, United Naitons, New York. (1982)
3. UNSCEAR Ionizing Radiaiton:Sources and biological effects. United Nations Scientific Commitee on the Effects of Atomic Radiaiton, United Naitons, New York. (1988)
4. Gür. F, and Günseli. Y, Journal of Environmental Science and Health, Part A. **45**, 1900-1908 (2010)
5. Bem, H., Wieczorrkowski, P., Budzanowski, M, J. Environ. Radioact. **61**, 191-201 (2002)
6. Aycik, G.A, and Ercan, A, J. Environ. Radioact. **35**, 23-35 (1997)
7. Akyüz. T, Akyüz. S, Varinlioğlu. A, Köse. A, Eric. J, Davies. D, Spectroscopy Letters. **29**, 1131-1139 (1996)
8. Turhan. Ş, Arıkan. I.H, Yücel. B, Varinlioğlu. A., Köse. A, Fuel. **89**, 2528-2535 (2010)
9. Beretka. J, Mathew. P.J, Health Phys. **48**, 87-95 (1985)
10. Kurnaz. A, Gezelge. M, Hançerlioğulları. A, Çetiner. M.A, Turhan. Ş, Human and Ecological Risk Assessment: An International Journal. **22**, 1375-1385 (2016)
11. UNSCEAR (2000) Ionizing Radiaiton: Sources and biological effects. United Nations Scientific Commitee on the Effects of Atomic Radiaiton, United Naitons, New York.

Natural radioelement concentrations in the soil of the Mila region of Algeria

Amina BRAMKI[1]*, *Mourad* RAMDHANE[2], *Fatima* BENRACHI[1]

[1]Mathematics and Subatomic Physics Laboratory (LMPS), University of Constantine-1, Constantine 25017, Algeria
[2]LPSC, Université Grenoble Alpes, CNRS/IN2P3, Institut National Polytechnique de Grenoble, F-38026 Grenoble Cedex, France

Abstract. In this study, the levels of the natural and artificial radioactivity in soil samples collected at various depths from Algerian agricultural region El-Athmania Mila was measured. Activity concentrations of the concerned radionuclides were determined by gamma-ray spectrometry using a high-purity germanium detector. The activity concentrations of ^{226}Ra, ^{232}Th and ^{40}K were found unchanged as a function of depth and ranged from 23.72±2.37to 65.47±5.06 Bq.kg^{-1} for ^{226}Ra, from 26.45±0.78 to 27.10±0.80 Bq.kg^{-1} for ^{232}Th and from 220.80±10.01 to 260.70±8.24 Bq.kg^{-1} for ^{40}K respectively. To evaluate the radiological hazard of radioactivity in samples, the radium equivalent activity (Raeq), the absorbed dose rate (D), the annual effective dose and the external (Hex) and internal hazard indices (Hin) were calculated. The mean of the excess lifetime cancer risk observed in this study are under the world's mean values.

1. Introduction

Every day, we ingest and inhale radionuclides via the consumption of food, water and air. All types of food including wheat and apple contain detectable amount of radioactivity, which successively relocate into the human body via the ingestion pathway. We know also, that the activity of food is strictly linked to the activity of soil where food was implanted. At present, studies on health effects due to ionizing radiations have produced substantial evidences that exposure to high level of radiations can cause illness or even death. Despite a well-known effect of cancer, scientists have long known that ionizing radiations with high doses may also cause mental retardations in the children of mothers exposed to radiations during pregnancy period [1].

Among the various implanted soils, the soil of wheat and apples has been selected to study their activity, because wheat and apples crops are very widespread. The natural radionuclides present in soil and fertilizers consist mainly of ^{238}U, ^{232}Th isotopes with their daughter products and the ^{40}K [2].

The knowledge of activity concentrations and distributions of the radionuclides in these materials are of greater interest since it provides useful information in the monitoring of environmental radioactivity [3].

The main objective of this paper is to determine the activity concentrations of natural and artificial radionuclides in the soil samples collected from the region of Mila and to evaluate the radiological hazard. The results, may be used, as a reference data for monitoring possible radioactivity pollutions in future.

2. Experimental procedures

2.1 Study area

Soil samples were collected from Mila region in the northeast of Algeria. Mila province lies in the longitude of 6° 16' E and in the altitude of 36° 27' N. It covers an area of 9375 km^2 and has a population of 768419. Mila among the best producers of cereals across the national territory.

2.2 Samples collection and preparation

In order to measure the natural radioactivity in the soil planted with apples and wheat, six soil samples were collected in the area referred in section 2.1. We took samples of soil planted with wheat and apples, at different depths going from 10 to 50 cm. The distance between the sites is 100 meters, separated by breezes Winds. The materials used for soil sampling are shovel, plastic bucket, boxes, pen or marker. The soil was collected in a clean plastic bucket and mixed well. Samples of 0.5 kg were ground, homogenized and sieved. They were dried for 48 h to ensure that the moisture is completely removed. Each sample was placed in a plastic box of radius 50 mm and 14 mm thick. Before the gamma analysis was carried out, the boxes had remained hermetically sealed for four weeks, in order to establish the radioactive secular equilibrium between the radionuclides of short half-lives. This step was necessary to ensure that the radon gas remains confined within the volume and that the daughters will remain in the sample [4].

2.3 Experimental setup

The measurement had been carried out in the laboratory of low activities in Grenoble (LBA/LPSC) equipped with two low noise HPGe detectors. Each detector is surrounded by two cm of archaeological lead exempt from natural radioactivity and 15 cm of purified lead. The two

* Corresponding author: amina.bramki@umc.edu.dz

detectors and their shielding is positioned at the center of a cube of two meters square of side surfaces, each face of the cube, with the exception of the ground, is in fact a liquid scintillation detector. These detectors act as veto prohibiting the acquisition of data during the passage of a cosmic ray. The energy calibration of the detectors was done using standard multi gamma sources. The resolution of the detectors in these measures was 0.85 keV and 1.85 keV at energies 122 keV and 1332 keV, respectively. The IAEA-RGU-1 and IAEA-RGTh-1 in powder form, installed in the same plastic box used for the samples, were used to determine the detectors efficiency curves. The activities of natural-series were 4940±30 Bq/kg for ^{238}U and 228±2 Bq/kg for ^{235}U in IAEA-RGU-1 and 3250±10 Bq/kg for ^{232}Th in IAEA-RGTh-1. InterWinner software (ITECH INSTRUMENTS) was used for the acquisition and treatment of the data collected by the detectors, taking into account the dead time. Each sample placed directly over the front of the detector was measured during an accumulating time between 48 and 72 h. The background spectrum was measured under the same conditions of samples during 1 month and was used to correct the calculated samples activities.

3. Calculations of activity concentrations and radiological indices

3.1 Activity concentrations

The activity concentrations of the radionuclides in the measured six samples (As) were computed using the following equation [5],

$$A_s(Bq.kg^{-1}) = \frac{N}{\varepsilon * I * m * t} \quad (1)$$

Where N is the net gamma counting, ε is the detector efficiency of the specific γ-ray, I is the absolute gamma intensity of gamma-decay, m is the masse of the sample (kg) and t is the time of measurements

3.2 Radiological Hazard Indices

3.2.1 *Radium equivalent activity*

As seen in the introductory part of this paper, the equivalent radium activity (Ra_{eq}) is a parameter used to estimate the external exposure due to the contribution of gamma emitters from ^{226}Ra, ^{232}Th and ^{40}K. Eq. (2) was employed for the calculation of the Ra_{eq}.

$$Ra_{eq} = A_{Ra} + 1.43A_{Th} + 0.077A_K \quad (2)$$

Where A_{Ra}, A_{Th} and A_K are, respectively, the radiological concentrations or specific activities obtained for ^{226}Ra, ^{232}Th and ^{40}K, using $Bq.kg^{-1}$ as a measurement unit [6]. The weights are based on the estimation that 370 $Bq.kg^{-1}$ of ^{226}Ra, 259 $Bq.kg^{-1}$ of ^{232}Th and 4810 $Bq.kg^{-1}$ of ^{40}K produce the same γ-ray dose [7].

3.2.2 *Hazard Indices (H)*

In order to evaluate the hazard of the natural γ radiation, the external hazard indices (Hex) was calculated from the following formula [8],

$$H_{ex} = \frac{A_{Ra}}{370\,Bq/Kg} + \frac{A_{Th}}{259\,Bq/Kg} + \frac{A_K}{4810\,Bq/Kg} < 1 \quad (3)$$

Where A_{Ra}, A_{Th}, and A_K are the activity concentrations of ^{226}Ra, ^{232}Th and ^{40}K. The Internal hazard indices (H_{in}) is given by Eq (4) [9],

$$H_{in} = \frac{A_{Ra}}{185\,Bq/Kg} + \frac{A_{Th}}{259\,Bq/Kg} + \frac{A_K}{4810\,Bq/Kg} < 1 \quad (4)$$

3.2.3 *Representative level indice*

Another radiation hazard index called the representative level indices ($I_{\gamma r}$) is defined as follows in Eq (5)

$$I_{gr} = \frac{A_{Ra}}{150\,Bq/Kg} + \frac{A_{Th}}{100\,Bq/Kg} + \frac{A_K}{1500\,Bq/Kg} \quad (5)$$

Where A_{Ra}, A_{Th} and A_K are specific activities of ^{226}Ra, ^{232}Th and ^{40}K in $Bq.kg^{-1}$, respectively [10].

3.2.4 *Absorbed dose rate*

The external gamma dose rate $\dot{D}$ (nGy h^{-1}) in air at 1 m above the ground level caused by the presence of natural radionuclides was calculated according to the formula UNSCEAR. 2008 [11] ,

$$\dot{D} = 0.462\, C_u + 0.604C_{Th} + 0.0417\, C_k \quad (6)$$

Where C_{Ra}, C_{Th} and C_K are the activity concentrations of ^{226}Ra, ^{232}Th and ^{40}K ($Bq.kg^{-1}$).

The annual effective dose equivalent and excess lifetime cancer risk

To estimate the annual effective dose rates, the conversion coefficient from absorbed dose in air to effective dose for adults is (c=0.7 Sv/Gy) and outdoor occupancy factor (p=0.2) proposed by UNSCEAR were used UNSCEAR 2000 [12]. The effective dose rate in unit of (μSv/y) was calculated by following equation:

$$D_E = c.\dot{D}.t.p.10^{-3} \quad (7)$$

Where t is the annual exposure time (8760 h). Excess lifetime cancer risk outdoors (presented in table 2) was calculated as follows [13]:

$$ELCR = D_E.D_L.PC \quad (8)$$

Where D_E (μSv) is defined in Eq (7), D_L is the duration of life (70 y) and PC is the nominal probability coefficient for detriment adjusted cancer risk of $5.5*10^{-2}$ Sv^{-1} for the whole population (ICRP. 2007).

4. Results and discussion

4.1 THE ACTIVITY CONCENTRATIONS

The results of activity concentrations in the soil samples from the six different sites are gives in table1 for the natural radionuclides of ^{226}Ra, ^{232}Th and ^{40}K and the artificial radionuclide ^{137}Cs. Radium equivalent activity (Ra_{eq}), representative level indices, external hazard indices, absorbed dose rates, annual effective doses and the excess lifetime risks of cancer in soil samples are given in table 2.

Table 1: Activity concentrations (Bq.kg^{-1}) of ^{226}Ra, ^{232}Th, ^{40}K and ^{137}Cs for each of the measured samples

S	B10	B30	B50	P10	P30	P50
^{40}K	220.80±10.01	238.00±9.44	248.6±12.58	254.00±10.30	260.70±8.24	257.10±8.30
^{226}Ra	23.72±2.37	46.00±6.89	65.47±5.06	46.14±8.35	55.83±5.80	43.10±7.50
^{232}Th	26.60±0.88	26.45±0.78	27.09±1.11	26.56±0.84	27.10±0.80	26.60±0.80
^{235}U	2.69±0.60	0.90±0.38	1.80±0.32	1.31±0.47	0.93±0.32	1.44±0.40
^{137}Cs	3.00±0.29	2.39±0.22	2.36±0.47	4.10±0.27	3.72±0.20	3.17±0.18

Table 2. The external (H_{ex}) and internal (H_{in}) hazard indices, radium equivalent (Ra_{eq}), representative level ($I_{\gamma r}$), air-absorbed dose rates ($\dot{D}$), annual effective doses (D_E) and excess lifetime cancer risk (ELCR) for samples.

Sample name	H_{ex} (Bq.kg^{-1})	H_{in} (Bq.kg^{-1})	Ra_{eq} (Bq.kg^{-1})	$I_{\gamma r}$	$\dot{D}$ (nGy.h^{-1})	D_E (μSv/y)	ELCR (10^{-4})
B10	0.21±0.04	0.28±0.05	78.8±14.0	0.57±0.10	37.40±6.67	45.87±8.18	1.76±0.31
B30	0.28±0.06	0.40±0.09	102.1±22.4	0.73±0.16	37.82±8.29	46.38±10.16	1.78±0.39
B50	0.33±0.06	0.51±0.09	123.4±20.8	0.87±0.15	44.86±7.58	55.02±9.29	2.11±0.35
P10	0.28±0.07	0.40±0.10	103.7±26.4	0.74±0.19	38.00±9.00	46.60±11.84	1.79±0.45
P30	0.31±0.05	0.46±0.08	114.7±18.9	0.82±0.13	40.47±6.68	49.63±8.19	1.91±0.31
P50	0.27±0.06	0.39±0.09	100.9±23.9	0.72±0.17	39.34±9.30	48.25±11.40	1.85±0.43

B10: Soil of wheat depth between 10 to 20 cm.
B30: Soil of wheat depth between 20 to 30 cm.
B50: Soil of wheat depth between 30 to 50 cm.
P10: Soil of apple depth between 10 to 20 cm.
P30: Soil of apple depth between 20 to 30 cm.
P50: Soil of apple depth between 30 to 50 cm.

The concentrations found in the present study ranged from 23.72±2.37 (B10) to 65.47±5.06 (B50) Bq/kg for ^{226}Ra, from 26.45±0.78 (B30) to 27.10±0.80 (P30) Bq/kg for ^{232}Th and from 220.80±10.01 (B10) to 260.70±8.24 for ^{40}K.

The average activity concentrations of ^{226}Ra, ^{232}Th and ^{40}K were found to be 46.71, 26.73 and 246.53 Bq/kg in soil samples, respectively. The world's mean values of ^{226}Ra, ^{232}Th and ^{40}K activity concentrations are 32, 45 and 420 Bq/Kg, respectively (5). The mean value of ^{226}Ra is higher than the world's average value.

We see that only for samples cultivated with wheat, concentrations of ^{40}K and ^{226}Ra change with depth. For all other samples, the concentration of radionuclides does not really change vertically.

For ^{235}U, the average activity concentration in samples was 1.51±0.42 Bq.kg^{-1}. This indicates a low presence of ^{235}U in samples because the natural abundance of ^{235}U is only 0.72% of the total uranium content [14].

The ^{232}Th activity is lower than that of ^{226}Ra, whereas the ^{40}K activity is the highest one and represents the largest contribution to the total activity of soil.

Our results show the presence of ^{137}Cs in the soil with low concentrations in all samples (mean value =3.12 Bq.kg^{-1}). Nuclear accidents, bomb tests and nuclear weapons are the origins of ^{137}Cs contamination. It has a half-life of about 30 years, it can be traced for several years.

The mean values of activity concentrations of ^{226}Ra, ^{232}Th and ^{40}K in soil samples from the studied area were compared with those from similar investigations in other countries such as the mean values in Jordan in 2014 are 57.7, 18.1,138.1 Bq.Kg^{-1} respectively [15]. They are however comparable to those determined in Algeria in 2011 with reported a mean values of 53.2, 50.0, 311.0 Bq.Kg^{-1} respectively.

The resultant values from the studied area are lower than the observed from reported in Algeria in 2011 but are higher from reported in Jordan in 2014 [16].

4.2 RADIOLOGICAL HAZARD INDICES

Table 2 shows the calculated values of radiological effects for different samples. The calculated values of the H_{ex} indices varied from 0.21 to 0.33, with a mean value of 0.28. For H_{in} the range was 0.28 – 0.51, with a mean value of 0.41. For all samples, H_{in} was less than 1

The values of calculated Ra_{eq} for the studied samples, are shown in the fourth column of Table 2. The calculated Ra_{eq} values range from 78.8 (B10) to 123.4 (B50) Bq.kg^{-1}. All the values of Ra_{eq} in the studied samples are lower than the critical limit of 370 Bq.kg^{-1} [10].

The representative level indices $I_{\gamma r}$ for all samples under study was smaller than unity. The calculated effective doses for the samples are under the worldwide value 80 μSv/y.

Finally, the excess lifetime cancer risk (ELCR) in the samples is ranged from 1.76x10^{-4} in B10 to 2.11x10^{-4} in B50 for adults. The samples are under the worldwide average of 2.54x10^{-4} (taking into account the average annual effective dose of 66 μSv from [11].

5. Conclusion

Activity concentrations of ^{226}Ra, ^{235}U, ^{232}Th, ^{40}K and ^{137}Cs were determined by using gamma-ray spectroscopy for soil samples at different depths from the El Athmania agricultural region of Mila in Algeria. The results show the presence of a low concentration of ^{137}Cs in soil, with a mean value of 3.12 Bq.kg^{-1}. Activity concentration of a naturally occurring radionuclide in samples were within the world average ranges which are 46.71 (10–50), 26.73 (7–50) and 246.53 (100– 700) Bq.kg^{-1} for ^{226}Ra, ^{232}Th and ^{40}K respectively [12].

The value of Ra_{eq} activity concentrations for soil samples is less than 370 Bq kg^{-1}.The average annual effective dosages are also lower than the global average values. ELCR varies from 1.76x10^{-4} to 2.11×10^{-4} with mean value of 1.87×10^{-4}. This study established a map of baseline information for future studies on radiation levels and radionuclide distribution in the environment of Mila. The results of the study serve as a reference for future assessment.

References

1. M. Rafik, S. Ur Rahman , M. Basharat c , W. Aziz , I. Ahmad , K. A. Lone , K.Ahmad and Matiullah, J. Radiat. Res. Appl. Sci. **7**, 29 (2014).
2. N.K. Ahmed, A.G.M. El-Arabi, J. Environ. Radioact. **84**, 51 (2005).
3. O. Agar, I. Boztosun, M. E. Korkmaz, S. F. Özmen, Rad. Prot. Dosimet. **162**, 630 (2014).
4. Oktay Baykara, Mahmut Doğru, Radiat. Meas. 41, 362 (2005).
5. Ch. Lakehal, M. Ramdhane, A. Boucenna, J. Environ. Radioact. **101**, 377 (2010).
6. J.A . Santos Júnior, R.S. Amaral, C.M. Silva, R.S.C. Menezes, Radiat. Meas. **45**, 861 (2010).
7. I.P. Farai, J.A. Ademola, J. Environ. Radioact. **79**, 119 (2005).
8. R. Ravisankar, A. Chandrasekaran, P. Vijayagopal, B. Venkatraman, G. Senthilkumar, P. Eswaran, A. Rajalakshmi, Radiat. Phys. Chem. **81**, 1789 (2012).
9. B.A. Almayahi, A.A. Tajuddin, M.S. Jaafar, Appli. Radia. Isotop. **70**, 2652 (2012).
10. NEA-OECD, 1979. Exposure to radiation from natural radioactivity in building materaials. Report by NEA group of Experts of the nuclear energy agency, OECD, Paris, France.
11. United Nations Scientific Committee on the Exposure of the Public and Workers from Various Sources of Radiation. UNSCEAR United Nations, New York (2008).
12. United Nations Scientific Committee on the Effects of Atomic Radiation. UNSCEAR United Nations, New York (2000).
13. M. Stajic, B. Milenkovic, M. Pucarevic, N. Stojic, I. Vasiljevic, D. Nikezic, Chemosph. **146**, 68 (2016).
14. A.M. El-Arabi. Radiat. Meas. **42**, 94 (2007).
15. H. Saleh . M. Abu Shayeb, Ann. Nucl. Energy . **65**, 184 (2014).
16. w. Boukhenfouf, A. boucenna, J. Environ. Radioact. **102**, 336 (2011).

Surface dose measurement with Gafchromic EBT3 film for intensity modulated radiotherapy technique

Ugur Akbas[1,*], *Nazmiye* Donmez Kesen[1], *Canan* Koksal[1], *Murat* Okutan[1], *Bayram* Demir[2], *Hatice Bilge* Becerir[1]

[1]Istanbul University, Oncology Institute, Department of Medical Physics, Istanbul, Turkey
[2]Istanbul University, Science Faculty, Department of Physics, Istanbul, Turkey

Abstract. Accurate dose measurement in the buildup region is extremely difficult. Studies have reported that treatment planning systems (TPS) cannot calculate surface dose accurately. The aim of the study was to compare the film measurements and TPS calculations for surface dose in head and neck cancer treatment using intensity modulated radiation therapy (IMRT). IMRT plans were generated for 5 head and neck cancer patients by using Varian Eclipse TPS. Quality assurance (QA) plans of these IMRT plans were created on rando phantoms for surface dose measurements. EBT3 films were cut in size of 2.5 x 2.5 cm^2 and placed on the left side, right side and the center of larynx and then the films were irradiated with 6 MV photon beams. The measured doses were compared with TPS. The results of TPS calculations were found to be lower compared to the EBT3 film measurements at all selected points. The lack of surface dose calculation in TPS should be considered while evaluating the radiotherapy plans.

1 Introduction

The dose deposited between air and phantom is defined as the surface dose. The dose at the surface has two components; contaminant electrons from the air, the linac collimator, the scattering materials in the path of the beam and secondary electrons that arise from irradiated patient [1]. Surface dose measurement is a challenging issue, due to steep dose reduction in surface area because of the absence of electron equilibrium. This steep dose gradient is causing requirement of special measurement techniques with appropriate dosimetric devices.

The most accurate dosimetric device for surface dose measurement is extrapolation ionization chamber, yet it is not owned in all institutes. Instead of extrapolation chamber, parallel plate ionization chambers can be used after making Gerbi's overresponse corrections [2]. The overresponse occurs because of the scattering electrons from the sidewall of the parallel plate chamber. The physical properties of these chambers confine the usage with only phantom measurements.

Radiochromic film is suitable dosimetry for surface dose measurements with their high spatial resolution and low spectral sensitivity. Bilge et al. [3] used radiochromic EBT film and Markus parallel plate ionization chamber in their surface dose measurements; they found the difference between EBT film and ionization chamber to be within 5% for 6MV and 3% for 18MV. Radiochromic films can be used as a good alternative to parallel plate chambers.

The treatment planning systems help us to receive the calculated doses at the related areas that irradiated with treatment plans. The prediction of surface dose with TPSs depends on some factors such as calculation algorithm, linac commissioning and beam modelling. Previous studies reported that TPSs still have not enough accuracy for surface dose calculations.

The aim of the study was to compare the film measurements and TPS calculations for surface dose in head and neck cancer treatment using intensity modulated radiation therapy.

2 Materials and Methods

Computed tomography (CT) images of 5 larynx cancer patients were acquired in head gantry-supine position. Thickness of the CT images was chosen 3 mm. Gross tumor volume, clinical target volume, planning target volume and organs at risk were defined and contoured by radiation oncologist. Then images were sent to TPS. Treatment plans were created by using Eclipse 8.9 (Varian, Palo Alto, CA) TPS on Varian Trilogy linac equipped with a 120-leaf MLC, utilizing 6 MV coplanar photon beams. Dose calculation was performed with Analytical Anisotropic Algorithm (AAA). Calculation grid size was selected 2.5 mm. IMRT plans were generated with 7 coplanar fields which were separated at 40° apart. Sliding window technique was chosen for all plans and a fixed dose rate of 500 MU/min was applied for each field. QA plans were created using rando

*Corresponding author: u.akbas@yahoo.com

DOI: 10.1051/epjconf/201715401011

phantom for irradiation in same conditions. Gantry, collimator and couch positions set to same positions with treatment plans.

GafChromic EBT3 film (International Specialty Product, NJ, US) was utilized. The EBT3 film has a single active layer that sandwiched between two transparent polyester sheets. The EBT3 film is more sensitive than the older versions of radiochromic film with its wide dose range from 1 cGy to 40 Gy [4,5]. A calibration curve of film batch was created before irradiation (Fig. 1). The EBT3 film was prepared into size of 2.5 × 2.5 cm^2 and put perpendicularly between the RW3 water equivalent slab phantoms at the depth of 5 cm where the linac was calibrated with 1 cGy equals to 1MU. The irradiation of EBT3 films were made with doses from 0–800 cGy at a field size of 10 x 10 cm^2. One piece of EBT3 film was not irradiated to use as a background.

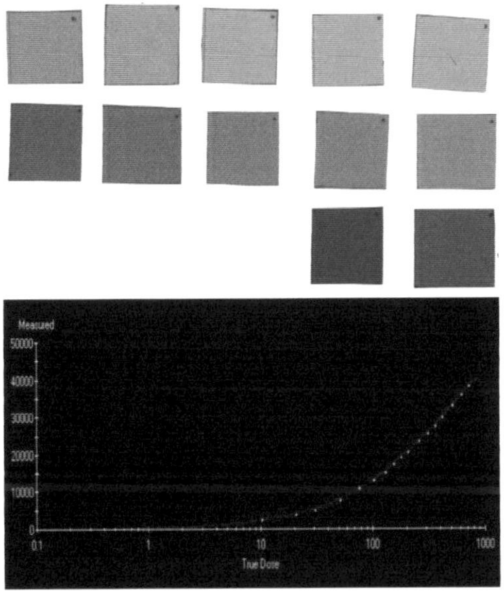

Fig. 1. The pieces of EBT3 film and the calibration curve of film batch

EBT3 films placed on the left side, right side and the center of larynx (Fig. 2). The films were irradiated with 6 MV photon beams. The exposed films were scanned after 24 hours from irradiation and then the calibration curve of the film batch was used to obtain the absolute doses. The measured doses were compared with TPS.

In this study, the comparison of film measurements and TPS calculations were also performed in 10 x 10 cm^2 open field using RW3 water equivalent slab phantoms for 6 MV photon beams to evaluate surface and buildup region doses in different setup conditions (Fig. 3).

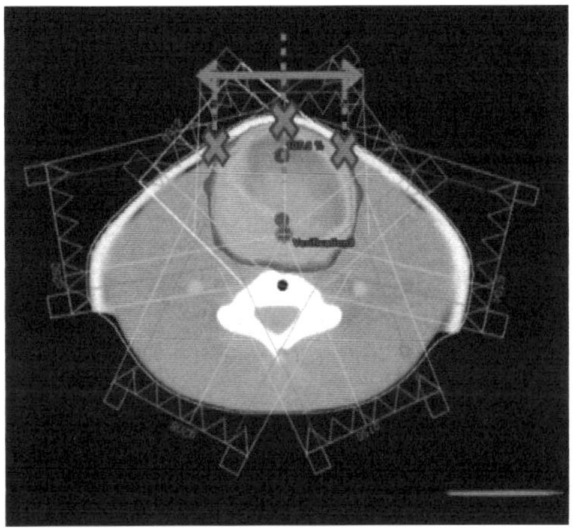

Fig. 2. Measurement points on rando phantom

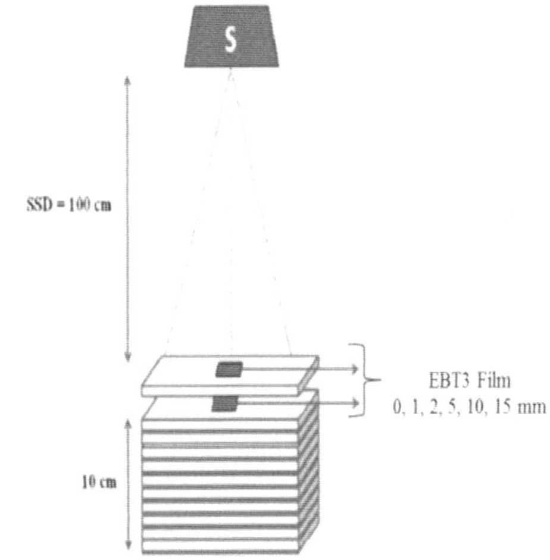

Fig. 3. Setup condition for open field irradiation

3 Results

The results of Gafchromic EBT3 film measurements and Varian Eclipse TPS calculations for 5 IMRT QA plans with 15 points (3 points for each) are shown in Fig.4. The surface doses were found to be higher with Gafchromic EBT3 film compared to Eclipse TPS for all points. The mean differences between measurements and calculations for 1st, 2nd, 3rd, 4th and 5th IMRT plans were 29.3%, 25.4%, 20.8%, 29.9% and 23.6%, respectively.

As mentioned before, in this study film measurements and TPS calculations compared in 10x10 cm^2 open field. The difference of surface and buildup region doses of EBT3 film and TPS in open field is shown in Table 1.

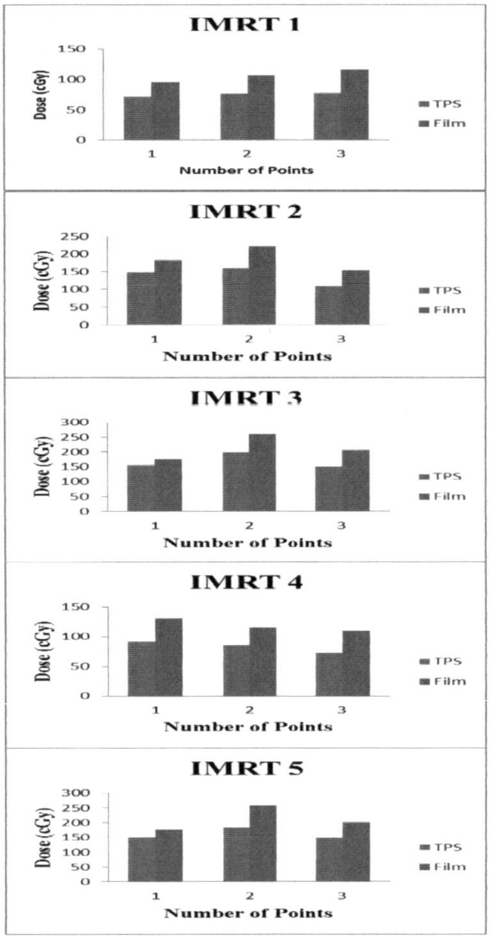

Fig. 4. The results of surface dose measurement and calculation for IMRT

Table 1. %Depth dose (%DD) values for open field

Depth (cm)	Gafchromic EBT3 Film (%DD)	TPS (%DD)
0	20.4	11.0
1	48.0	42.2
2	61.3	54.5
5	87.6	79.9
10	99.2	95.8
15	100.0	100.0

4 Conclusion

Radiotherapy is an important treatment modality for the head and neck cancer. One of the most common intensity modulated technique, IMRT, has advantage of delivering maximum dose to target while sparing the healthy tissues. Immobilization, CT scanning, contouring the volumes of targets and normal tissues, technique of treatment planning, set-up conditions and dose delivery are the components of treatment. The dose deposited at surface in radiotherapy is very important in cases where the patient skin is defined as dose limiting tissue or part of the target in the irradiation area. An overexposure of the skin may give rise to acute reactions of the skin or the delayed effects of normal tissues.

The surface doses were found to be lower in TPS calculation compared to EBT3 film measurements both in IMRT and open field irradiations. The surface dose difference between TPS calculation and EBT3 measurement is greater in IMRT because of the using oblique fields. The studies reported that the surface dose increases with oblique angles; the relative dose might be 50% larger at an obliquity of ~55° [6, 7].

Akino et al. [8] found that the measured dose with film dosimetry at 3 mm depth was higher than TPS calculation results by 15-30% for all treatment techniques but at deeper depths TPS calculation and film measurements results were found to be close. Devic et al. [4] reported that the even small thickness of EBT film can increase surface dose because of the steep dose gradient in the buildup region and should be accounted during calibration process. Chow et al. [9] compared the calculation accuracy of a superposition/convolution algorithm for oblique beams with Monte Carlo calculation, and reported that both the AAA and collapsed cone algorithms cannot accurately calculate dose distribution at depths less than 2 mm.

Accurate knowledge of the dose at surface can be helpful to prevent the acute reactions and delayed effects. Our study has agreement with previous studies about the deficiency of TPS calculation in shallow depths. The lack of surface dose calculation in TPS should be considered while evaluating the radiotherapy plans.

References

1. A. Sigamani et al., Journal of Medical Physics/Association of Medical Physicists of India. 2016;**41**(2):85-91.
2. D. E. Mellenberg Jr. Medical Physics. 1990;**17**(6):1041–1044.
3. H. Bilge, A. Cakir, M. Okutan, H. Acar. Physica Medica. 2009;**25**(2):101–104.
4. S. Devic, et al., Medical Physics. 2006;**33**(4):1116–1124.
5. L.J. van Battum, D. Hoffmans, H. Piersma, S. Heukelom. Medical Physics, 2008;**35**(2):704–716.
6. F.M. Khan, *The Physics of Radiation Therapy*, Lippincott Williams &Wilkins, Philadelphia, USA, 4th edition, 2010.
7. J.-P. Lin, T.-C. Chu, S.-Y. Lin, M.-T. Liu. Applied Radiation and Isotopes. 2001;**55**(3):383–391.
8. Y. Akino, I.J. Das, G. K. Bartlett, H. Zhang, E. Thompson, J.E. Zook. Medical Physics. 2013;**40**(1):011714 – 6.
9. J. C. Chow, R. Jiang, M.K. Leung. Journal of Applied Clinical Medical Physics. 2010;**12**(1):3424.

Nuclear structure of In isotopes in ^{100}Sn mass region

Nadjet Laouet*, and *Fatima* Benrachi

LPMS Laboratory, Frères Mentouri Constantine 1 University, 25000 Constantine, Algeria

Abstract. The monopole effect resulting from the interaction between the magic core and the valence particles has a particular interest in the study of nuclear structure. To understand the importance of this interaction, we have realized some spectroscopic calculations for odd-odd In isotopes containing one hole proton and few neutron particles in addition to ^{100}Sn doubly magic core in their valence spaces. The using interaction is derived from jj45apn one taking into account the monopole interaction in the studied mass region, and using recent single particle and hole energies. The calculations are performed in the framework of the nuclear shell model by means of Oxbash nuclear structure code.

1 Introduction

The region around the last magic N=Z nucleus ^{100}Sn, close to the path of rp-process, was the main of several theoretical and experimental works that aimed to give a global description of nuclear structure. With their one proton hole and few particle neutrons, Odd-Odd indium isotopes, near ^{100}Sn doubly magic core, are of great importance in nuclear structure studies. They give opportunity to develop our knowledge about proton-neutron interaction near astrophysical processes pathways.

^{102}In was identified for the first time by Béraud et al. [1,2]. They studied the A=102 isobars using an on-line mass isotopic separator operating on nuclear reactions induced by heavy ions. This isotope was produced by means of the reaction ^{92}Mo (^{14}N, 4n), and the half-life of 24(4) s is measured. Martín et al. [3] have investigated nuclei in end-point region of the rp-process near ^{100}Sn using direct mass measurements at SHIPTRAP by the Penning trap mass spectrometer at GSI Darmstadt, where they experimentally identified three nuclides for the first time. The mass excess of ^{102}In is determined to be -70694(12) keV, which is different by 16 keV from the given value Audi et al. [4]. In 2009, Elomaa et al. [5] have realised a mass filter of nuclei produced via proton and ^{3}H induced fusion-evaporation reactions in ^{100}Sn region. They have measured a new mass excess -70690.4(54) keV which is different by 4 keV from the previous one. This nucleus was the subject of a theoretical study in Ref. [6]. In which, Karny et al. have performed spectroscopic calculations of the β-decay of ^{102}Sn within the $\pi(2p_{1/2},1g_{9/2})$ $\nu(1g_{7/2}, 2d_{5/2}, 2d_{3/2}, 3s_{1/2}, 1h_{11/2})$ model space, and using SNC interaction [7]. In comparison with the experimental data, this interaction has reproduced the spin and parity of the ground state 6^+ and the three first excited states. However, the calculated energies for 2_1^+ and 3_2^+ states are far from the experimental ones.

In 1977, Varley et al. presented the discovery of ^{104}In [2,8]. They have realised measurement of half-lives, excitation functions, γ-x-ray and γ–γ coincidences in order to determine the energetic spectrum of ^{104}In. The measured half-life of the isomer was estimated to be 1.5(0.2) min. Szerypo et al. [9] studied the beta decay of ^{104}Sn at GSI, by means of ^{58}Ni+^{50}Cr reaction in order to evaluate the energetic spectrum of the ^{104}In descendent. Few years later, Karny et al. [6] have studied the beta decays of $^{102\text{-}104}$Sn and their $^{102\text{-}104}$In descendents using high-resolution germanium detectors as well as a Total Absorption Spectrometer (TAS). Elomaa et al. [5] have measured the mass excess of ^{104}In -76176.5(51) keV with a difference of 66 keV from the given one in [4]. The experimental spectra of these two isotopes are presented Fig.1.

2 Theoretical framework

The interactions the supposed inert core with the adding nucleons can lead to shell evolution and modification of the spectroscopic properties of nuclei near around this core [10-12].

* Corresponding author: nadjet.laouet@umc.edu.dz

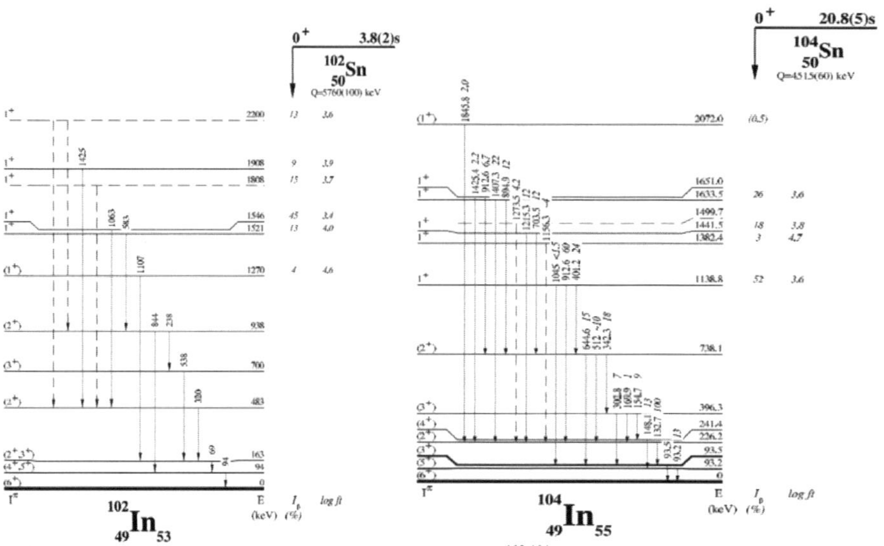

Fig.1. Experimental spectra of $^{102-104}$In [6, 9].

Poves and Zuker [13] have introduced the description of the monopole effect, in which the Hamiltonian of the system is defined in terms of the two-body interaction. Hence, the consideration of this effect can reproduce the missing nuclear properties of nuclei far from stability.

They have proposed to express the monopole Hamiltonian of the system in terms°: [14-16],

$$H_m = \sum_s n_s \varepsilon_s + \sum_{s \le t} \left(a_{st} n_{st} + b_{st} T_{st} \right) \qquad (1)$$

s and/or t denote a proton and/or a neutron orbit. $n_{s,\,t}$ and $T_{s,\,t}$ refer, respectively, to the number and the isospin operator defined by Zuker [10,15] as a function of the monopole Hamiltonian diagonal part $V_{st}^{\tau\tau'}$ [12]. This amount can be defined as a function of the average two body matrix elements (*TBMEs*), of a given effective interaction V_J $(j_s j_t)$, over the configurations of s and t orbits [12,14],

$$V_{st}^{\tau\tau'} = \frac{\sum_J (2J+1) V_J (j_s j_t)}{\sum_J (2J+1)} \qquad (2)$$

Here, τ (τ') stands for proton or neutron.

In this work, we have used the recent single particle energies (*SPEs*), and considered the mass and the monopole effects to introduce some modifications on the two body matrix elements (*TBMEs*) of the original interaction *jj45apn* from ^{78}Ni mass region Jensen [17, 18]). These TBMEs are used in order to calculate the monopole term (Eq. 2): $V^{pn}_{1g_{9/2}2d_{5/2}} \approx -430\,keV$ to modify $\pi\upsilon(1g_{9/2}\,2d_{5/2})^{T=0}_{J=2,7}$ *TBMEs*. These *TBMEs* are chosen basing on the energetic sequence of the single particle space.

Using the resulting interaction *jj45m* and the original one, some calculations are carried out in order to reproduce the nuclear properties of $^{102-104}In$ isotopes.

3 Results and discussion

In this work, we have performed shell model calculations, by means of the new interaction *jj45m* in $\pi(0f^{-1}_{5/2}, 1p^{-1}_{3/2}, 1p^{-1}_{1/2} \text{ and } 0g^{-1}_{9/2})^{Z-28}$ and $\nu(0g_{7/2}$, $1d_{5/2}$, $1d_{3/2}$, $2s_{1/2}$ and $1h_{11/2})^{N-50}$ model space in ^{100}Sn magic core. The experimental single hole and single particle energies taken, respectively, from ^{99}In for protons and ^{101}Sn for neutrons are used as a starting point to calculate the effective single particle energies [19, 20].

The calculations for ^{102}In nucleus using *jj45m* interaction, shown in Fig. 2, allow to reproduce the parity of the low laying states. However, this interaction gives 7^+ as a ground state, which is different from 6+, the experimental one. The original interaction gives negative parity for the low laying states.

Neither the energetic excited states, nor the experimental sequence are reproduced using the two interactions for ^{104}In nucleus. However, the original interaction lead to reproduce the levels sequence except the 6^+ state, which is situated between 4^+ and 8^+ states.

The reduced electromagnetic transition probabilities can be expressed in terms of the electromagnetic *TBMEs* $\langle J_f \| M_{\sigma\lambda} \| J_i \rangle$:

$$B(M_{\sigma\lambda} : J_i \to J_f) = \frac{1}{2J_i + 1} \left| \langle J_f \| M_{\sigma\lambda} \| J_i \rangle \right|^2 \qquad (3)$$

Table 1 and 2 show the electromagnetic properties evaluated by means of *jj45apn* and *jj45m* interactions, for the studied nuclei.

The calculated values of reduced electric transition probabilities *B(E2)* are obtained using $e_p = 1.50e$ and $e_n = 0.50e$ for the effective charges. For the magnetic moments μ_J, we have used the free g factors. These results show that the two interactions give different values for the electromagnetic properties. However, the obtaining mean lives are close for ^{104}In.

DOI: 10.1051/epjconf/201715401012

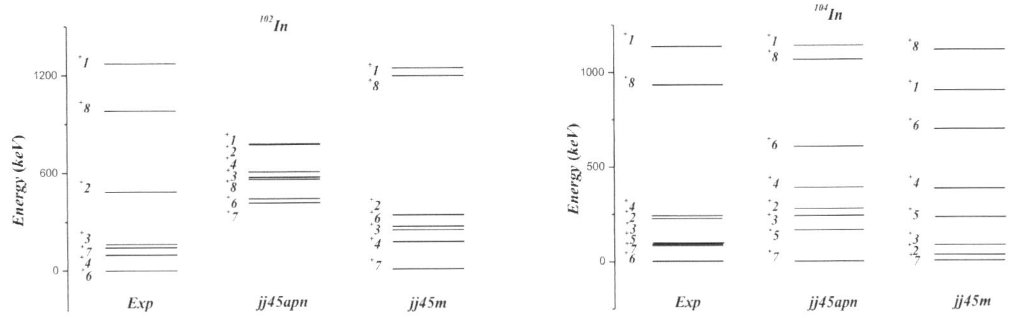

Fig. 1. Calculated energetic spectra for $^{102\text{-}104}In$ isotope using *jj45apn* [17] and *jj45m* interactions in comparison with the experimental data [21].

Table 1. Electromagnetic properties and mean lives calculated using *jj45apn* and *jj45m* interactions, for ^{102}In.

J_i	J_f	$B(E2: e^2fm^4)$		$B(M1: \mu_N^2)$		$t(ps)$		$\mu(Ji: \mu_N)$		$Q(Ji:efm^2)$	
		jj45apn	*jj45m*	*jj45apn*	*jj45m*	*jj45apn*	*jj45m*	*jj45apn*	*jj45m*	*jj45apn*	*jj45m*
1^+	2^+	0.959	3.641	3.792	0.039	$0.937\ 10^5$	1.963	0.620	0.604	-8.990	-8.783
	3^+	25.100	1.836	/	/						
2^+	3^+	2.393	90.570	0.726	2.145	0.950	32.910	1.664	3.797	-8.998	8.844
	4^+	0.181	13.250	/	/						
4^+	3^+	35.480	104.144	0.184	1.579	62.690	$3.048\ 10^3$	5.124	5.026	6.980	6.441
	5^+	127.100	129.300	0.872	1.749						
	6^+	4.902	30.434	/	/						
5^+	6^+	35.570	30.916	0.242	0.834	984.200	$5.016\ 10^5$	5.913	5.723	12.080	8.253

Table 2. Electromagnetic properties and mean lives calculated using *jj45apn* and *jj45m* interactions, for ^{104}In.

J_i	J_f	$B(E2: e^2fm^4)$		$B(M1: \mu_N^2)$		$t(ps)$		$\mu(Ji: \mu_N)$		$Q(Ji:efm^2)$	
		jj45apn	*jj45m*	*jj45apn*	*jj45m*	*jj45apn*	*jj45m*	*jj45apn*	*jj45m*	*jj45apn*	*jj45m*
1^+	2^+	9.209	34.680	0.471	1.363	0.189	0.064	-0.099	3.079	-1.265	3.777
	3^+	1.485	1.234	/	/						
2^+	3^+	178.100	188.72	1.365	1.723	650.400		4.236	4.135	15.750	14.050
	4^+	44.046	61.812	/	/						
4^+	3^+	60.220	79.800	0.859	1.096	2.553	1.609	4.373	4.631	0.113	1.042
	5^+	78.150	82.870	1.716	1.809						
	6^+	37.570	52.173	/	/						
5^+	6^+	38.504	53.548	1.798	1.096	$2.856\ 10^5$	$0.431\ 10^5$	5.424	5.546	9.873	9.581

Conclusion

This study is based on the energetic spectra and electromagnetic properties calculations, for odd- odd Indium isotopes, with one hole proton and few neutrons in their valence spaces. The calculations are carried out in the framework of the nuclear shell model, by means of *Oxbash* nuclear structure code. Using the jj45apn original interaction of the code, we realized some modifications based on the mass effect and proton-neutron monopole interaction to get *jj45m* one. The new interaction lead to reproduce the parity of the first excited states for ^{102}In isotope. However, it can't reproduce neither the sequence nor the excited states for ^{104}In nucleus. The two interactions give different results for the electromagnetic properties and the mean lives are close for ^{104}In. These results are the consequence of the differences on the obtained energetic spectra for both nuclei.

Authors of this article thanks to the organizers of TESNAT 2017 May 10th-12th 2017 Adana-Turkey for the organization and the support provided during the workshop. Special thanks are owed to B. A. Brown, for his help in providing us the Oxbash code (Windows Version), and to M. H. Jensen, for the documents and the information provided about the interaction jj45apn.

References

1. B. Béraud et al., Z. Phys. A **299**, 279 (1981)
2. S. Amos and M. Thoennessen, arxiv :nucl-ex/ 1004.5266v2 (2009)
3. A. Martin et al., Eur. Phys. J. A **34**, 341 (2007)
4. G. Audi, A.H. Wapstra and C. Thibault, Nucl. Phys. A **729**, 337 (2003).
5. V.V. Elomaa et al., Eur. Phys. J. A **9**, 1 (2009)
6. M. Karny et al., Eur. Phys. J. A **27**, 129 (2006)
7. B. A. Brown and K. Rykaczewski, Phys. Rev. C **50**, R2270 (1994)
8. B.J. Varley , J.C. Cunnane and W. Gelletly, J. Phys. G: Nucl. Phys. **3**, 55 (1977)
9. J. Szerypo et al., Nucl. Phys. A **507**, 357 (1990)
10. N. Smirnova et al., Phys. Lett. B **686**, 109 (2010)
11. O. Sorlin and M.G. Porquet, Prog. Part. Nucl. Phys., **61**, 602 (2008)
12. A. Umeya et al., Phys. Rev. C **77**, 034318 (2008)
13. A. Poves and A.P. Zuker, Phys. Rep. **70**, 235 (1981)
14. A.P. Zuker, Phys. Scr. T **88**, 157 (2000)
15. A.P. Zuker, Phys. Rev. Lett. **90**, 042502 (2003)
16. T. Otsuka et al., Phys. Rev. Lett. **105**, 032501 (2010)
17. M. Hjorth-Jensen, T.T.S. Kuo and E. Osnes, Phys. Rep. **261**, 125 (1995)
18. M. Rejmund et al., Phys. Rev. C **93**, 024312 (2016)
19. W. Wang et al., Chinese Phys. C **36**, 1603 (2012)
20. H. Grawe et al., Rep. Prog. Phys. **70**, 1525 (2007)
21. http://www.nndc.bnl.gov/chart/

Dosimetric comparison between VMAT and RC3D techniques: case of prostate treatment

Fatima Zohra Chemingui[1,*], *Fatima* Benrachi[1], *Mohamed Saleh* Bali[2], and *Hamid* Ladjal[3]

[1]LPMPS Laboratory, Physics Department, Frères Mentouri Constantine1 University, 2500 Constantine, Algeria
[2]Athena Medical Center, 25000 Constantine, Algeria
[3]LIRIS Laboratory, Claude Bernard university, 69100 Lyon, France

Abstract. Considered as the second men cancer in Algeria, prostate cancer is treated in 70% by radiation. That's why radiation therapy is therapeutic weapon for prostate cancer. Conformational Radiotherapy in 3D is the most common technique [1–5]. The use of conventionally optimized treatment plans was compared at case scenario of optimized treatment plans VMAT for prostate cancer. The evaluation of the two optimizations strategies focused on the resulting plans ability to retain dose objectives under the influence of patient set up. Dose Volume Histogram in the Planning Target Volume and dose in the Organs At Risks were used to calculate the conformity index, and evaluation ratio of irradiated volume which represent the main tool of comparison [6,7]. The situation was analysed systematically. The 14% dose increase in the target leads to a decrease in the dose in adjacent organs with 39% in the bladder. Therefore, the criterion for better efficacy and less toxicity reveal that VMAT is the best choice.

1 Introduction

Cancer is a term used when abnormal cells divide without control and are able to invade other tissues. In this case, cells can spread to other parts of the body through the blood and lymphatic systems. It is a major topic worldwide. Considered to be responsible for high percentage deaths in Algeria, prostate cancer is the biggest people killer aged between 60-80 years. The treatment of cancer is summarized in three main categories: drugs, surgery and radiation-therapy. The choice of the treatment depends on the tumor type, location, stage of the disease and general state of patient. Most of time, the treatment is combined. Its ultimate goal is to eradicate the disease, without damaging the surrounding healthy tissues. In practice, the medical decision is always a trade-off and compromise between Tumor Control Probability (TCP) and Normal Tissue Complication Probability (NTCP) [3, 5, 8–10].

At present, radiation therapy (RT) is used in about 70% of all cancer treatments in industrialized countries. It allows killing tumor cells, by deposing a high dose of radiation within the tumor. In North Africa and exactly in Algeria the first private Anti Cancer Center "Athena Medical Center", treating with conventional techniques, was opened in 2013. The evolution of photon delivery techniques using conventional linear accelerator equipped with multi-leaf collimator as RC3D Conformational or VMAT [11–15]. The main task of a medical physicist is to make sure that the right dose is delivered in the right place. Treatment planning systems (TPS) was used to plan radio therapy treatments, but with limited dose accuracy in some particular cases. On the other hand, the OAR must be conserved and receive the minimum of dose. In prostate cancer, the better ballistic of VMAT over RC3D allows decreasing the dose to surrounding tissues and achieves through 76 Gy in PTV.

2 Material and method

Treatment planning (TP) is the function that translates the physician's prescription and intent into a treatment plan. It contains the set of equipment parameters that control the dose deposition in patient expected to be equivalent to the treatment plan predicted dose. This definition often implies many requirements often above and beyond those implied for photon TP. For the optimization, the physician's prescription is a set of quantified statements (i.e., maximum dose to target), suitable for numeric manipulation in an optimization algorithm, and generalized intent (i.e., minimize dose to the OAR). The quality of a particular objective is limited by the prescription that fixes the boundaries of what remains possible in a trade-off consideration of competing objectives [11,16,17].

* Corresponding author: z.chemingui@hotmail.fr

 DOI: 10.1051/epjconf/201715401013

In this context, two patients with curator treatment were the subject of this retrospective study. They were selected for stage and volume. The treatment parameters for a chosen patient are listed in Table 1. The two patients have undergone 3DCT simulation using immobilization device (headrest, footrest, foot wedge medical solutions) with General Electric Optima Radiation Therapy CT scanner. Scans covered the entire pelvis volume and upper abdomen part. Then, CT data imaging were generated using the treatment planning system Eclipse (soma version 11.3 by Varian Medical System, California, USA). Gross tumor volume (GTV) was contoured by the treating radiation oncologist from CT datasets. For each prostate cancer case two treatment plans were generated accounting for treatment uncertainties in two different ways:

VMAT plans were developed using the Eclipse (Version 11.3 Varian Medical System, California, USA) Treatment Planning System (TPS) with 6 and 18 MV for each patient. AAA (Analytical Anisotropic Algorithm, Varian Medical System, California, USA) was used to compute the dose distributions. Inverse pelvis volume and upper abdomen part. Then, CT data imaging were generated using the treatment planning system Eclipse (soma version 11.3 by Varian Medical System, California, USA). Gross tumor volume (GTV) was contoured by the treating radiation oncologist from CT datasets. For each prostate cancer case two treatment plans were generated accounting for treatment uncertainties in two different ways:

VMAT plans were developed using the Eclipse (Version 11.3 Varian Medical System, California, USA) TPS with 6 and 18 MV for each patient. AAA (Analytical Anisotropic Algorithm, Varian Medical System, California, USA) was used to compute the dose distributions. Inverse treatment plans for VMAT were generated using the same dose-volume constraints for all plans. The dose constraints was set for the rectal wall, rectum, bladder, bowel, femoral heads and unspecified normal structure.

For the RC3D radiotherapy planning, after contouring all normal structures and critical organ, 7fields treatment techniques were used for the smallest volume PTV 76Gy. Beam arrangements 0°, 45°, 90°, 135°, 225°, 270°, 315° angles or orientations were used. The prescribed dose was normalized to 100% at the isocenter, and 95% isodose surface covered the PTV.

Dose Volume Histograms (DVHs) were used to compare treatment plans including PTV, OAR, between two treatment techniques. Then, doses distributions were evaluated and in particular points of DVHs [12] by conformity index (CI) defined by tumoral volume covered by reference Isodose, (CI, ideal value = 1),

$$CI = \frac{V_{IR}}{V_T} \quad (1)$$

V_{IR} and V_T denote respectively volume of reference Isodose, and Tumoral Volume.

3 Results and discussion

Fig 1. shows DVHs curves of dose distribution in prostatic irradiation with 76Gy.

Dose distribution illustrated by DVHs curves for patient treated for prostate cancer: (Right) VMAT plan, (left) RC3D plan. The PTV 76Gy is represented in red, in Dark brown: PTV 46Gy, in Blue: left femoral head, in Cyan: Right Femoral head, in Green: Bowel, in Brown: Rectum, Orange: Bladder, Pink: Prostate.

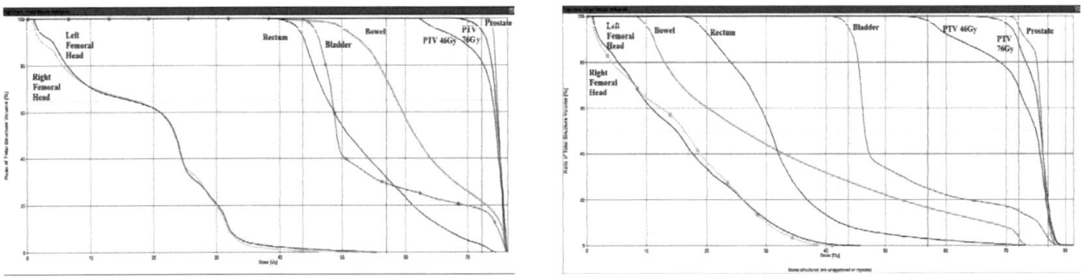

Fig. 1. DVHs: Comparison of dose distribution between VMAT and RC3D

Dose differences between VMAT and RC3D plans in different structures of OAR (rectum wall, rectum, bladder, bowel, and femoral heads) are illustrated in Table 1. This observational study shows DVHs Curves of prostatic irradiation at 76Gy, for VMAT and RC3D plans. The mean DVH are obtained in OARs, which are mentioned above. The DVHs curves obtained by VMAT are underneath those obtained by RC3D. For the other structures; bladder and soft tissues, the curves cross each other in several times. We can observe that the mean dose in different OARs is significantly reduced in VMAT plans. The values of this reduction attend their maximum for the bladder with more than 40% of reduction in the mean dose.

Table 1. Difference in the delivered mean dose and maximum dose between the RC3D plan and the VMAT plan for PTVs and OARs

Organs	Volume (cm^3)	ΔDose max (Gy)	ΔDose mean (Gy)
Rectum	74	-1,132	-21,597
Rectal wall	30,2	-6,771	-23,422
Bladder	142,8	-1,230	-30,797
Bowel	1094,7	-4,455	-4,646
FBR	174,6	-9,619	-4,343
FBL	181,8	-11,508	-3,792
PTV 60Gy	190,1	5,056	21,085
PTV 76Gy	141,4	5,056	0,643

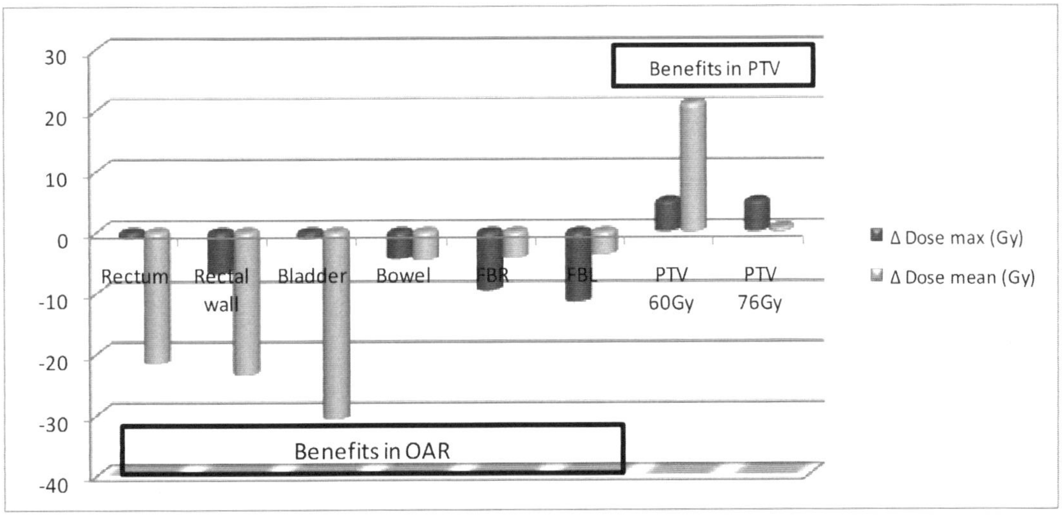

Fig. 2. Graphical representation of Δ dose means and max between the VMAT and RC3D plans

Table 2. Comparison between Conformity indexes of VMAT and 3DRC plans

	CI RC3D ISODOSE 95% 76GY (72.2)	CI VMAT ISODOSE 95% 76GY (72.2)	Diff CI
Patient 1	0.7312	1.0855	0.3543
Patient 2	0.7956	0.9306	0.135

In the other hand, we can see in Table 2 the difference in term of CI between VMAT and RC3D plans. The values of CI for the RC3D plans are less than 0.8 otherwise VMAT plan values are upper than 0.9.

Dose values (maxima and mean) of particular PTV 46 Gy, PTV 60 Gy, PTV 76 Gy and OARs are exposed, and we can observe that the mean dose in rectum decreases by about 30% with VMAT technique. In volume covered by dose of 60 Gy (V60) and 70 Gy (V70), it decreases respectively by 20%, 14%. For bladder, it decreases by 10% and 18% respectively. For femoral heads, VMAT dose decrease in V30 from 20% to 9%. Table 2 shows VMAT technique conformity index values, varied from 0.93 to 1.0 and are all close to the ideal value otherwise for RC3D technique.

To help the analysis, we affected some statistical tests, and we calculated the VMAT improvement ratio of conformation in PTVs. the results are represented in Fig. 3.

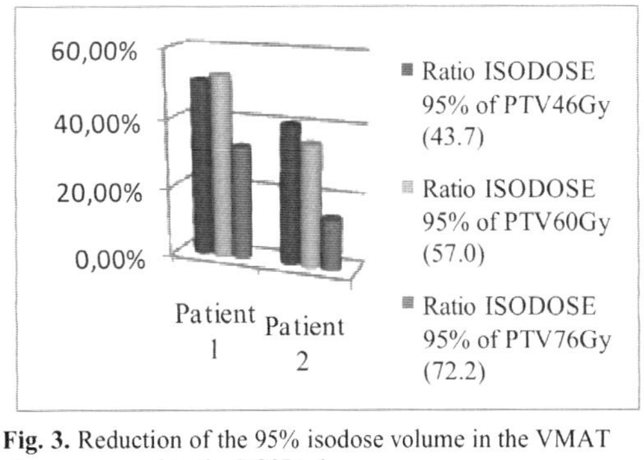

Fig. 3. Reduction of the 95% isodose volume in the VMAT plans compared to the RC3D plans.

The PTV 46 Gy volume is ameliorated by factor of 50% due to volume containing all OARs. For PTV 60 Gy, the improvement is more than 30%, this volume is smaller than PTV 46 Gy it contains rectum and bladder.

For PTV 76 Gy, the value of this improvement is more than 14% and consists prostate only.

Conclusion

In this study, we evaluated the dose distribution plans obtained from two techniques RC3D and VMAT. The VMAT plans gave more effectiveness in treated volumes by increasing the dose mean in PTVs corresponding to the prostate tumor and the nodal regions. VMAT technique offers best conformation in target volumes (PTVs) and less toxicity in soft tissues (less dose in OARs). Actually, in developed countries VMAT have replaced RC3D in the treatment of prostate cancer. In Algeria, Athena Medical Center is the only cancer center offering this technique in clinical routine.

References

1. F. Alongi, et al., Rad. Oncol. **8**, 171 (2013)
2. D. Wolff et al., Med. Phys. German. **19**, 58 (2009)
3. M. M. Aspradakis et al., Phys. Med. Biol. **48**, 2873 (2003)
4. R. Garcia, Validations dosimétriques des conditions cliniques des radiothérapies thoraciques, (Ph. D Thesis, Université Toulouse III - Paul Sabatier 2013)
5. ICRU Report 78, J. ICRU 7, (2007)
6. C. Popescu et al., Int. J. Rad. Onco. Biol. Phys. **76**, 287 (2009)
7. J. Zhu, Modèles prédictifs de toxicité en radiothérapie par modulation d'intensité, (Ph. D Thesis, Université de Rennes I. 2013)
8. ESTRO booklet n 09: Guidelines for the verification of IMRT (2008)
9. A. Pollack et al., Rad. Onco. Biol. Phys. **53**, 1097 (2002)
10. H. M. Kooy et al., Philadelphia, Lippincott. Williams. & Wilkins. **70**, 70 (2008)
11. D. M. Hall et al., Rad. Onco. **10**, 115 (2016)
12. J. He et al., J. Thorac. Dis. **8**, 862 (2016)
13. A. Cakir et al., Asian. Pac. J. Cancer. Prev. **16**, 3365 (2015)
14. C. Vargas, Int. J. Rad. Onco. Biol. Phys. **62**, 1297 (2005)
15. S.J. Feigenberg et al., Int. J. Onco. Biol. Phys. **62**, 397 (2005)
16. M. Valeriani et al., Biol. Med. Res. Int. **10**, 6 (2014)
17. P.A. Kupelian *et al.*, *Int J Rad. Onco. Biol. Phys.* **58**, 25 (2004)

S_4 solution of the transport equation for eigenvalues using Legendre polynomials

Hakan Öztürk[*], and *Ahmet* Bülbül

Osmaniye Korkut Ata University, Faculty of Arts and Sciences, Department of Physics, Osmaniye, Turkey

Abstract. Numerical solution of the transport equation for monoenergetic neutrons scattered isotropically through the medium of a finite homogeneous slab is studied for the determination of the eigenvalues. After obtaining the discrete ordinates form of the transport equation, separated homogeneous and particular solutions are formed and then the eigenvalues are calculated using the Gauss-Legendre quadrature set. Then, the calculated eigenvalues for various values of the c_0, the mean number of secondary neutrons per collision, are given in the tables.

1 Introduction

As well known, the number of neutrons and their distributions are very important in order to maintain the constant power in all nuclear power plants. In other words, the multiplication factor which describes the ratio of the produced number of fission neutrons in one generation to the number of neutrons in previous generation is desired to be equal to one. The stability of the power produced in the reactor can be thought as the isotropic scattering of the neutrons through the system.
There is no analytic solution of the transport equation which describes the interactions and the distribution of the neutrons with the materials inside the system. The methods developed for the solution of the transport equation can be interpreted as deterministic (such as polynomial expansion based techniques, integral transform methods) and stochastic (such as Monte Carlo, MCNP, source iteration). In a stochastic approach, the transport equation is usually converted into a discrete ordinate form (S_N) [1-3].

In this study, first the neutron transport equation is converted into a discrete ordinates form and then the S_N transport equation is solved numerically for the eigenvalue spectrum of the mono-energetic neutrons. The results are obtained up to the fourth order approximation (S_4) using Legendre quadrature set and they are given in the table for various values of the c, number of secondary neutrons per collision.

2 Method

The neutron transport equation for mono-energetic neutrons in a slab with isotropic scattering can be written as,

$$\mu\frac{\partial\psi(x,\mu)}{\partial x}+\sigma_T\psi(x,\mu)=\frac{\sigma_{S0}}{2}\int_{-1}^{1}\psi(x,\mu')\,\mathrm{d}\mu'+\frac{Q_0}{2}, \quad (1)$$

where $\psi(x,\mu)$ is the neutron angular flux at position x travelling in direction μ, cosine of the angle between the neutron velocity vector and the positive x-axis. σ_T is the total macroscopic cross-section, σ_{S0} is the differential scattering cross-section corresponding to isotropic scattering and Q_0 is the internal source [3]. When the integral in Eq. (1) is written as an integral transform with Legendre polynomials, it can be rearranged in the form of discrete ordinates S_N equations for the numerical solution,

$$\mu_m\frac{\mathrm{d}\psi_m(x)}{\mathrm{d}x}+\sigma_T\psi_m(x)=\frac{\sigma_{S0}}{2}\sum_{n=1}^{N}\psi_n(x)\omega_n+\frac{Q_0(x)}{2}, \quad (2)$$

where ω_n is the Gauss-Legendre quadrature weights or weighting factor for direction μ_n, i.e. the roots of the Nth order Legendre polynomials. The roots or the zeros of the Legendre polynomials together with the weighting factors $\omega_m=1$, one can easily analyze Eq. (2) for the solution.

Therefore, the general solution of Eq. (2) can be written as the sum of the homogeneous and the particular solutions,

$$\psi_m(x)=\psi_m^p(x)+\psi_m^h(x). \quad (3)$$

It is easy to verify a spatially constant particular solution of the form can be given as,

$$\psi_m^p(x)=\frac{Q_0}{2\sigma_T(1-c_0)}, \quad 0\le a\le x, \quad 1\le m\le N, \quad (4)$$

[*] Corresponding author: hakanozturk@osmaniye.edu.tr

where $c_0 = \sigma_{S0} / \sigma_T$. In order to determine the homogeneous solution $\psi_m^h(x)$ of Eq. (2), the traditional method of separation of variables can be used. Therefore, the solution of homogeneous part of Eq. (2) can be written in the form of [3],

$$\psi_m^h(x) = H_m(\nu)\exp(\sigma_T x / \nu), \quad 0 \le a \le x, \ 1 \le m \le N. \quad (5)$$

By substituting Eq. (5) into the homogeneous part of Eq. (2), one can obtain an expression for the angular part of the neutron angular flux;

$$H_m(\nu) = \frac{\nu c_0}{2(\nu + \mu_m)} \sum_{n=1}^{N} H_n(\nu)\omega_n, \quad (6)$$

where the function $H_m(\nu)$ can be normalized as,

$$\sum_{n=1}^{N} H_n(\nu)\omega_n = 1. \quad (7)$$

Eq. (7) is multiplied by ω_m from both sides and then summed over all m to obtain an equation for the ν eigenvalues:

$$\frac{1}{2}\sum_{m=1}^{N} \frac{\nu c_0}{(\nu + \mu_m)} \omega_m = 1, \qquad \nu \ne -\mu_m. \quad (8)$$

Eq. (8) is referred to as the dispersion relation and the roots $\nu_k, 1 \le k \le N$, of Eq. (8) are the eigenvalues of the S_N equations and they are lying symmetrically about the origin for any c_0 satisfying $0 \le c_0 \le 1$.

3 Numerical results and discussion

The neutron transport equation for one-speed neutrons in a finite homogeneous slab is studied to determine the eigenvalues using S_N method with Gauss-Legendre quadrature set. The transport equation in one-dimensional geometry is first written in the form of S_N by using the integral transform technique with the even-order Gauss-Legendre quadrature set. A dispersion relation is obtained by using the reasonable homogeneous solution in the S_N form of the transport equation. The roots, i.e. the eigenvalues $\nu_k, 1 \le k \le N$ of Eq. (8) are calculated up to the forth order approximation for various values of c_0 [4,5]. This order of approximation can be spelled to as unnecessary for the calculations in slab geometry. However, this study should be thought as the first attempts for the problem solution strategy. This method could be approved for other problems in other geometries of the transport theory.

The calculated eigenvalues for various values of the $c_0 < 1$ and $c_0 > 1$ are given in Table 1. The calculation of the eigenvalues is accepted as to be the first step by many of the researchers. Therefore, the results presented here can also be used in many studies related with transport theory.

Table 1. Eigenvalue spectrum

N	$c_0 = 0.40$	$c_0 = 0.70$	$c_0 = 0.99$	$c_0 = 1.20$	$c_0 = 2.00$
2	±0.745355992	±1.054092553	±5.773502691	±1.290994449i*	±0.557735027i
4	±0.423258332	±0.492070474	±0.539816206	±0.562247603	±0.603465390
	±1.546700021	±0.940739334i	±0.547857547i	±0.440083387i	±0.280100324i

*$i = \sqrt{-1}$

References

1. E.W. Larsen, Nucl. Sci. Eng. **82**, 47 (1982)
2. Los Alamos National Laboratory, Physics Division Progress Report, Page 70 (1995)
3. K.M. Case, P.F. Zweifel, *Linear Transport Theory* (Addison-Wesley Company, 1967)
4. F. Scheid, *Numerical Analysis* (McGraw Hill, 1988)
5. F. Anli, Ann. Nucl. Energy, **28**, 1033 (2001)

DOI: 10.1051/epjconf/201715401015

Natural radioactivity investigation in Dam sediments of northeast Algeria using gamma spectroscopy

Fatima Benrachi[1*], *Ghania* Bouhila[1], *Asma* Saadi[1], and *Mourad* Ramdhane[2]

[1]LPMS Laboratory, Frères Mentouri Constantine 1 University, 25000 Constantine, Algeria
[2] LPSC, Université Grenoble Alpes, CNRS/IN2P3, Institut Nationale Polytechnique de Grenoble, F-38026 Grenoble Cedex France

Abstract. Current research paper intends to estimate the natural radioactivity levels in sediments samples collected from Beni Haroun Dam in the northeast Algeria, using high resolution HPGe detector. The mean activity concentrations values measured for the radionuclides ^{232}Th, ^{226}Ra and ^{40}K are 18.9 ± 1.9, 37.3 ± 2.7 and 149.9 ± 5.5 Bq/kg, respectively. The ^{137}Cs anthropogenic radionuclide has been observed with maximum activity concentration value of 0.8 ± 0.4 Bq/kg, which is considered an insignificant amount. In order to assess the radiological threat of gamma radiations emitted by these radionuclides on the health of the population, absorbed dose rate, annual effective dose equivalent and radiation hazard indices were had been calculated. The obtained values are compared with the world wide average ones.

1 Introduction

Every living creature in the environment is exposed to ionizing radiation comes from natural and anthropogenic radionuclides. Most of natural radioactivity present in environment arises from radioactive isotopes of ^{40}K, ^{238}U and ^{232}Th decay series [1,2]. The specific levels of natural environmental radiation are related to the composition of each lithologically area, and to the contents of the rocks from which the sediments or soils originate [3]. Furthermore, radiations originating from anthropogenic activities like nuclear weapons testing, diagnostic and therapeutic medicine, and radiation from nuclear plants can also released large amounts of radionuclides such ^{137}Cs in environment [4]. Natural sources make up approximately 80% of the total radiation dose which a person is exposed to in a year, while 20% come from anthropogenic sources [5]. Thus, the determination of distributions of the terrestrial and anthropogenic radionuclides in soil and sediment is of great interest for many researchers in the world [2-4]. In this work, the study area was Beni Haroun Dam, which located at 36° 33' 19'' N 6° 16' 11'' E, North-Eastern of Algeria. The aim of this study is to evaluate the ^{238}U, ^{232}Th, ^{40}K and ^{137}Cs radionuclide concentrations to provide background data on natural and artificial radioactive isotopes for the studied region, using gamma-ray spectrometry in low background configuration.

Additionally, an assessment of the radiological risks on humans arising from radioactivity sediment samples was determined in terms of the radiological parameters: the absorbed dose rate (D), the annual effective dose equivalent (AED) and radiation hazard indices were had been calculated. The obtained values are compared with the world wide average ones.

2 Materials and methods

2.1 Study area

Beni Haroun Dam, constructed in August 2003, is the largest dam in Algeria, located in northeast of Algeria, at about 15 km of the Mila city. The artificial dam was built at the confluences of Oued Rhumel and Oued Endja to 40 km north of Constantine Province. The Dam covers an area of 3.929 ha and its depth is 120 meters with a capacity of 960 million cubic meters of water. It supplies five provinces with potable as well as irrigation water [6, 7].

2.2 Sample collection and preparation

Five sediment samples were collected from different locations near Beni Haroun Dam, at a depth of 5-10 cm from the top surface layer as shown in Figure 1. Samples were crushed and dried at 110°C temperature in an oven until their weight become constant. Then, they were sieved to fine powder, passed through a standard 1*mm* mesh size and sealed into small cylindrical plastic containers having the same geometry as detector. Before measurements, the containers were stored approximately for 4 weeks (~ 7 half-lives) to reach secular equilibrium

* Corresponding author: fatima.benrachi@umc.edu.dz

between the ^{238}U and ^{232}Th series and their respective progeny [8, 9].

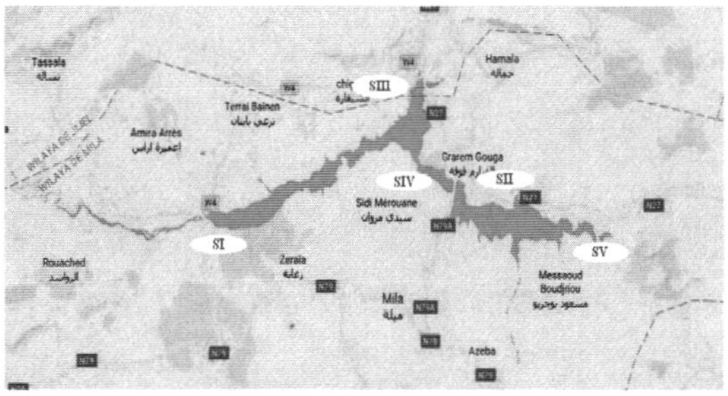

Fig. 1. Map of location samples

2.3 Gamma ray spectrometric analysis

The activity concentrations of samples were measured at low activity laboratory (LBA/LPSC-Grenoble) which has two HPGe detectors. Both detectors are surrounded by 2 cm of archeological lead and by 15 cm of purified lead. The two detectors and the lead plate are placed at the center of a two-meter-high cube; each face of this cube being constituted with a liquid scintillation detector. These detectors proceed as dynamic veto stopping the acquisition of data during the passage of a cosmic ray. The data acquisition and analysis use two PCs equipped with the "Interwinner" (ITECH-instruments) software. The detectors have a relative efficiency of 20% and the energy resolution was of 0.78 and 1.85 keV at 122 keV (^{57}Co) and 1332 keV (^{60}Co) gamma lines, respectively [10]. IAEA-RGU-1 and IAEA-RGTh-1, in powder form, installed in the same position and plastic bottle as the samples, were used to determine the detectors efficiency curves. The activities of the natural-series were 4940±30 Bq/kg for ^{238}U and 228±2 Bq/kg for ^{235}U in IAEA-RGU-1 and 3250±10 Bq/kg for ^{232}Th in IAEA-RGTh-1 [11]. Each sample was measured for 48 hours. The measured prominent gamma-ray lines from both the Uranium and Thorium decay series and ^{40}K, ^{137}Cs (Table 1) were used for sample analysis to determine the activity concentrations after correcting by background counts.

Table 1 : Prominent gamma energy used for estimation of radionuclides actvity

Radio nuclide	Energy (keV)	Gamma-ray intensity (%)	Radio Nuclide	Energy (keV)	Gamma -ray intensity (%)
^{238}U series			^{232}Th series		
^{234}Th	63.29	4.80	^{228}Ac	911.2	25.80
^{214}Pb	351.93	37.60	^{212}Pb	238.6	43.30
^{214}Bi	609.31	46.10	^{212}Bi	727.3	6.58
^{210}Pb	46.54	4.10	^{208}Tl	583.1	30.30
^{40}K	1460.8	10.72	^{137}Cs	661.6	85.10

3 Results and discussion

3.1 Estimation of activity concentrations

In The activity concentrations for the radionuclides in the five measured samples were evaluated using the relation:

$$A_s = \frac{N_E}{\varepsilon_E ptm} \tag{1}$$

Where: N_E and ε_E are the net peak area and the detection efficiency at energy E, p is the branching ratio of radionuclide of interest; t is the counting time (s); m is the mass of the measured sample in *kg*.

The results of the concentration values of activity ^{226}Ra, ^{232}Th, ^{40}K and ^{137}Cs reported in Figure 2 show an inhomogeneous distribution of the elements. This difference is probably due to the variation in the geological composition of the various sites near the dam from which the samples were taken. The ^{40}K is the largest contributor to the total activity of all samples. This is explained by the presence of the clay in all the places of the dam, because the clays contain a relatively high concentration of potassium. On the other hand, the activity of ^{226}Ra is higher than that of ^{232}Th in almost all samples, this could be related to the difference of their solubility in the natural environment. From this figure, the average activities of ^{226}Ra, ^{232}Th and ^{40}K obtained in the study are below the global average of 35, 33 and 400 Bq/kg as reported by UNSCEAR [5]. The ^{137}Cs anthropogenic radionuclide has been observed with maximum activity concentration value of 0.8 ± 0.4 Bq/kg, which is considered an insignificant amount.

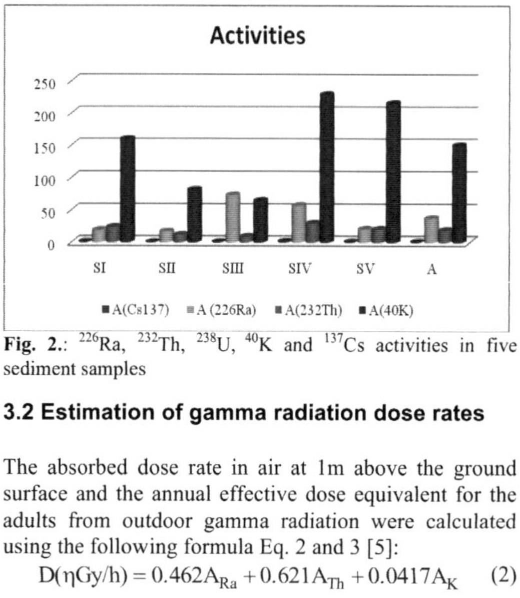

Fig. 2.: ^{226}Ra, ^{232}Th, ^{238}U, ^{40}K and ^{137}Cs activities in five sediment samples

3.2 Estimation of gamma radiation dose rates

The absorbed dose rate in air at 1m above the ground surface and the annual effective dose equivalent for the adults from outdoor gamma radiation were calculated using the following formula Eq. 2 and 3 [5]:

$$D(\eta Gy/h) = 0.462A_{Ra} + 0.621A_{Th} + 0.0417A_K \tag{2}$$

$$AED(\mu Sv/y) = D(\eta Gy/h) \times 8760h \times 0.2 \times 0.7 Sv/Gy \times 10^{-3} \tag{3}$$

Figure 3 and 4 show the calculation results of absorbed dose and annual effective dose equivalent, respectively. Their mean values were found to be 35 ηGyh^{-1} and 43 μsvy^{-1}, which are lower in comparison with the world average value, which equal to 59 ηGyh^{-1}, 70 μsvy^{-1}, respectively [5].

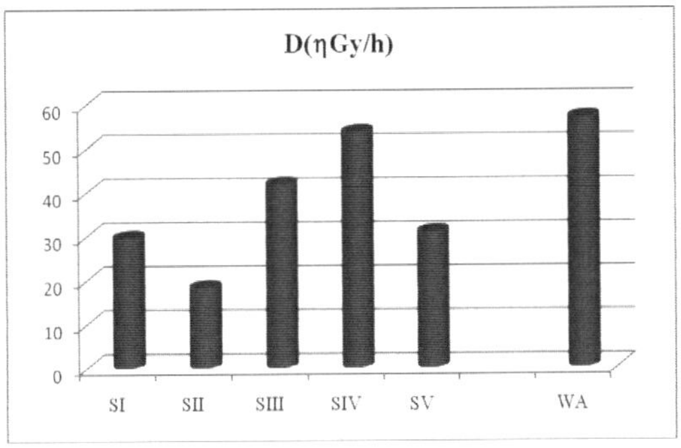

Fig. 3.: Absorbed dose rate of sediment samples

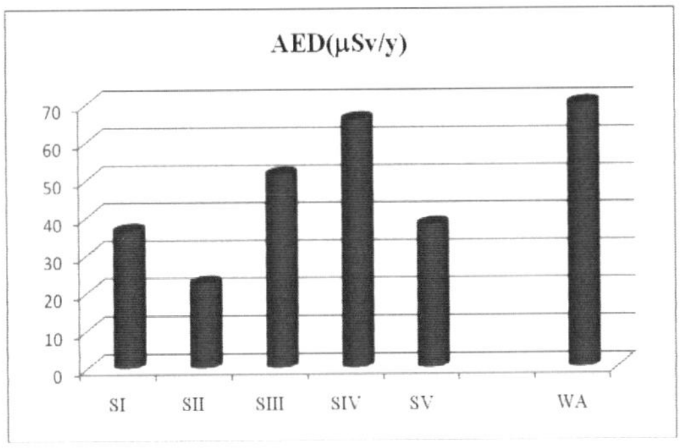

Fig. 4. Annual effective dose of sediment samples

3.3 Estimation of Radiation hazard indices

Besides, the results were evaluated in terms of the radiation hazard by means of the radium equivalent activity (Ra_{eq}) which is calculated through the relation given by [12].

$$Ra_{eq} = 0.077A_K + 1.43A_{Th} + A_{Ra} \quad (4)$$

The Ra_{eq} values for investigated locations are shown in Table 2. In addition, H_{ex} and H_{in} indices were calculated by means [12]:

$$H_{ex} = \frac{A_{Ra}}{370} + \frac{A_{Th}}{259} + \frac{A_K}{4810} \quad (5)$$

$$H_{in} = \frac{A_{Ra}}{185} + \frac{A_{Th}}{259} + \frac{A_K}{4810} \quad (6)$$

In the Table 2, the values of H_{ex} and H_{in} indices values range from 0.11 to 0.32 and 0.15 to 0.47 respectively. The Ra_{eq}, H_{ex} and H_{in} values in the study area are below than their permissible limits, which indicate that the sediments do not pose any significant radiation hazard.

Table 2. Radiation hazard indices for investigation samples.

Locations	Radiation hazard indexes		
	Ra_{eq}(Bq kg^{-1})	H_{ex}	H_{in}
SI	65.03±1.31	0.18±0.01	0.23±0.06
SII2	66.64±0.95	0.18±0.01	0.24±0.06
SIII	40.10±0.53	0.11±0.01	0.15±0.03
SIV	90.90±1.57	0.25±0.01	0.44±0.04
SV	116.95±1.45	0.32±0.01	0.47±0.05
Mean Value	75.93±3.90	0.21±0.21	0.31±0.25

Conclusion

Natural radioactivity concentrations of ^{226}Ra, ^{232}Th and ^{40}K have been measured in sediment samples collected from different locations around Beni haroun Dam in the northeast Algeria, using high resolution HPGe detector. Then, the gamma radiation dose rates values are estimated. The results indicate that the mean activity levels of radionuclides in all samples are within the world average values reported by UNSECEAR [5]. Moreover, the calculated values of absorbed dose rate, annual effective dose equivalent and radiation hazard indices were found under their permissible limits. The ^{137}Cs anthropogenic radionuclide has been observed with an insignificant amount. The present study was carried out to give a baseline reference data.

Authors of this article thanks the organizers of the 3rd International Conference on Theoretical and Experimental studies in nuclear applications and technology 10-12 May 2017 Adana-Turkey for the organization and the support provided during the conference.

References

1. X. Lu, X. Zhang, F. Wang, Environ. Geol. **53**, 1475 (2008).
2. A. Khatir Sam et al., Radiat. Prot. Dosim. **71(2)**, 141 (1997).
3. M. Mohery, Shadiah Baz, Adel M. Kelany, A.M. Abdallah. Rad. Phys. Chem. **97**, 16 (2014).
4. E.O. Agbalagba, G.O. Avwiri, Y.E. Chad-Umoreh. J. Environ. Rad. **109**, 64 (2012).
5. United Nations Scientific Committee on the Effects of Atomic Radiation (UNSCEAR), *Sources and Effects of Ionizing Radiation* (Report to the General Assembly) (New York: United Nation) (2000).
6. R. Marmi, M. Kacimi, M. Boularek, Revista de geomorfologie, **10**, 51 (2008).
7. A. Mebarki, *Ressources en eau et aménagement en Algérie. Le bassin du Kébir-Rhumel.* Alger, Office des Publications Universitaires (1984).
8. T. Santawamaitre et al., J. Environ. Rad. **138**, 80 (2014).
9. A. Nada et al., Appl. Radiat. Isot. **67**, 643 (2009).
10. http://lpsc.in2p3.fr.
11. *Preparation of gamma-ray spectrometry reference materials* RGU-1, RGTh-1 and RGK-1 report –AIEA /RL/**148**, Vienna, 1987
12. J. Beretka, P. J. Mathew, Health Phys. **48**, 87 (1985).

Spectroscopic properties of ^{130}Sb, ^{132}Te and ^{134}I nuclei in $^{100-132}$Sn magic cores

Fatima Benrachi*, *Meriem* Khiter, and *Nadjet* Laouet

LPMS Laboratory, Frères Mentouri Constantine 1 University, 25000 Constantine, Algeria

Abstract. We have performed shell model calculations by means of Oxbash nuclear structure code using recent experimental single particle (spes) and single hole (shes) energies with valence space models above the ^{100}sn and ^{132}sn doubly magic cores. The two-body matrix elements (tbme) of original CD-Bonn realistic interaction are introduced after have been modified taking into account the three-body forces. We have focused our study on spectroscopic properties evaluation of ^{130}Sb, ^{132}Te and ^{134}I nuclei, in particular their energy spectra, transition probabilities and moments have been determined. The getting spectra are in reasonable agreement with the experimental data.

1 Introduction

The effective interaction is a key ingredient for the success of the nuclear shell model. Once a reliable interaction is obtained, we can describe various nuclear properties accurately and systematically, which helps us to understand the underlying structure, and to make predictions for unobserved properties. Structure properties of some nuclei are important inputs for astrophysical r-process model calculations. Particularly, the study of nuclei in the vicinity of doubly magic Sn is indeed a subject of great current experimental and theoretical interest. Experimental information on these nuclei, which have been long inaccessible to spectroscopic studies, is becoming available.

In the last decades, many experimental studies were focused on the discovery of exotic nuclei. In 1948, the ^{134}I isotope was identified by Katcoff et al. [1,2]. Plutonium foils were irradiated with neutrons in the Los Alamos homogeneous pile Fission. In addition, excited levels up to an energy 3691 keV in ^{134}I were measured through the γ rays from the spontaneous fission of ^{252}Cf [3] and in the reaction ^{136}Xe+^{208}Pb [4]. In 2009, Coraggio et al. [5] have interpreted the five-observed high-excited states on the basis of a realistic shell-model calculation. They attributed spin-parity assignments to iodine states and they were found very good agreement between the experimental and calculated energies. Very recently, Laouet and Benrachi [6] have performed a study of odd-odd particle-hole nuclei in ^{132}Sn mass region basing on the effect of proton-neutron monopole interaction on the nuclear structure of N=81 isotones. The first identification of ^{132}Te was provided by Katcoff et al. in [1]. Besides, Brown et al. [7] have reported shell-model calculation of magnetic moments for selected Sn and Te isotopes, by means a residual interaction based on the CD-Bonn renormalized G-matrix. In a recent paper [8,9], two new levels above 10^+ isomer state have been experimentally identified in ^{132}Te. Hagebo et al. [10] reported the discovery of ^{130}Sb in 1962. Enriched ^{130}Te targets were irradiated with neutrons produced by bombarding beryllium with a 21 MeV deuteron beam from the Amsterdam cyclotron and ^{130}Sb was formed in (n,p) charge exchange reactions. Covello et al. [11] have performed shell-model calculations for nuclei close to doubly magic ^{132}Sn and ^{100}Sn focusing on particle-hole multiplets. They made realistic effective interactions derived from the CD-Bonn nucleon-nucleon potential. They have found that the agreement between experiment and theory is remarkably good for these calculations for ^{130}Sb. It should be noted that both the experimental and calculated energies are relative to the 8^- state, which has been experimentally observed to be the ground state.

2 Theoretical framework

The two body realistic interactions derived from the N-N force fail to reproduce some nuclear properties [12]. This problem can be solved by the consideration of the monopole interaction [13]. Zuker had shown in [13,14] that a very simple three body monopole term can solve practically all the spectroscopic problems in the p, sd, and fp shells. Those were assumed to need drastic revisions of the realistic two-body potentials. In 2010, Otsuka et al. [15] have also shown that three-body forces are necessary to explain drip line for oxygen isotopes. Therefore, Sarkar have incorporated [12] a simple three-body monopole term as prescribed by Zuker in [13,14]. Poves and Zuker [14] proposed to separate the Hamiltonian of the system into two parts:

$$H = H_m + H_M \tag{1}$$

where H_M and H_m denote respectively the multipole and the monopole Hamiltonians. The monopole Hamiltonian is defined as:

* Corresponding author: fatima.benrachi@umc.edu.dz

$$H_m = E_0 + \sum_i \varepsilon_i n_i + \sum_{i \le j} V_{ij} \frac{n_i(n_j - \delta_{ij})}{(1+\delta_{ij})}$$
$$V_{ij} = \frac{\sum_J (2J+1) V_{ijij}^J}{\sum_J (2J+1)} \qquad (2)$$

where ε_i corresponds the single particle energies (spes) and n_i , n_j are the valence nucleon numbers in the i and j shells, respectively.

For an inert core of closed shells (Z, N), the monopole term of the proton-neutron interaction can be determined by using differences in binding energies between neighboring nuclei and it given by [16],

$$V^{pn}_{j\pi j\nu} = \frac{1}{2}\left[BE(Z+1, N+2) - BE(Z, N+2)\right] - \left[BE(Z+1, N) - BE(Z, N)\right] \qquad (3)$$

The main aim of this paper is to present some nuclear properties from this study, focusing attention on the proton-neutron multiplets in the ^{130}Sb, ^{132}Te and ^{134}I nuclei in the framework of the nuclear shell model by means of Oxbash nuclear structure code [17].

2.1 Cores ^{100}Sn and ^{132}Sn

For calculations, the CD-Bonn in the proton-neutron formalism and Sn132ph interactions were used. The CD-Bonn interaction in the isospin formalism developed by Machleidt [18, 19] was based on the predictions of Bonn interaction for $j \le 4$ [18]. Basing on the mass dependence effect, we have modified the two body matrix elements (tbmes) of the original CD-Bonn [19] in order to perform shell model calculations using ^{132}Sn doubly magic nuclei as an inert core. To evaluate these modifications, we have used $\left(\frac{100}{132}\right)^{1/3}$ mass factor. The resulting two body matrix elements have been changed using the monopole interaction between $\pi g_{7/2}$ and $\nu d_{3/2}$ shells with $V^{pn}_{1g_{7/2}2d_{3/2}} \approx 150$keV to modify $(1g_{7/2}2d_{3/2})^{T=0}_{J=2-5}$ *tbmes*. The experimental proton spes taken from the spectra of ^{133}Sb and neutron shes taken from ^{131}Sn are used [20, 21] and new interaction named Sn132ph is introduced. The calculations have been realized in the jj55pn valence space. This later consists of five orbitals for both protons and neutrons above the ^{100}Sn inert core. The studied nuclei in this work have 30, 32 and 34 total valence particles with one, two and three proton particles, respectively. Most of the neutron orbitals are completely filled up, and these nuclei contain, respectively, three, two and one neutron holes. Therefore, no truncation was needed for the present calculation.

2.2 Probabilities and moments

The reduced electromagnetic transition probabilities are calculated by the form:

$$B(M_{\sigma\lambda}: J_i \to J_f) = (2J_i+1)^{-1} \left|\langle J_f \| M_{\sigma\lambda} \| J_i \rangle\right|^2 \qquad (4)$$

In the calculation of the reduced magnetic transition probabilities B(M1), we use the effective factors g^p_s (eff.)=3.79, g^n_s(eff.)= -2.04, g^p_l(eff.)=1.102 and g^n_l(eff.)= -0.005. These values are adopted for nuclei in ^{132}Sn mass region [22].

The lowest negative parity states of the studied nuclei are dominated by $\{\pi 1g_{7/2}, \nu 2d_{3/2}; J\}$ and configuration with $j_p = 7/2$ and $j_n = 3/2$. The electromagnetic moments of a state J with different proton and neutron spins can be calculated by means of [23]:

$$\mu_J = \mu_N \frac{J}{2}\left[\frac{\mu_j(p)}{j_p} + \frac{\mu_j(n)}{j_n} + \left(\frac{\mu_j(p)}{j_p} - \frac{\mu_j(n)}{j_n}\right)\frac{j_p(j_p+1) - j_n(j_n+1)}{J(J+1)}\right] \qquad (5)$$

$$Q_J = \begin{pmatrix} J & 2 & J \\ -J & 0 & J \end{pmatrix} (-1)^{j_p+j_n+J}(2J+1) \left[\begin{Bmatrix} j_p & J & j_n \\ J & j_p & 2 \end{Bmatrix} \frac{Q_{j_p}}{\begin{pmatrix} j_p & 2 & j_p \\ -j_p & 0 & j_p \end{pmatrix}} + \begin{Bmatrix} j_n & J & j_p \\ J & j_n & 2 \end{Bmatrix} \frac{Q_{j_n}}{\begin{pmatrix} j_n & 2 & j_n \\ -j_n & 0 & j_n \end{pmatrix}} \right] \qquad (6)$$

Here, $Q_{jp,\,jn}$ denote respectively single proton/neutron quadrupole moment

$$Q_j(p,n) = -\frac{(2j-1)}{(2j+2)} \langle r^2 \rangle e^{p,n}_{\lambda=2}(eff.) \qquad (7)$$

$e^{p,n}_{\lambda=2}(eff)$ is the effective charge of the nucleon and $\langle r^2 \rangle$ is the mean square radius of the nucleon in that orbital. Whereas, we have calculated the quadrupole electric moments using $e_p = 1.35e$ and $e_n = 0.9e$ effective charges.

3 Results and discussion

We have performed shell-model calculations be means the new interactions Sn132ph in jj55pn space model using ^{100}Sn and ^{132}Sn doubly magic cores. The experimental ground states of ^{130}Sb, ^{132}Te and ^{134}I nuclei in ^{132}Sn mass region are dominated by $\pi(1g_{7/2})^1\nu(1h_{11/2})^{-3}$, $\pi(1g_{7/2})^2\nu(2d_{3/2})^{-2}$ and $\pi(1g_{7/2})^3\nu(2d_{3/2s})^{-1}$, respectively. Our spectroscopic calculations gave the same configuration for the ground states except for ^{130}Sb.

Fig.1(a) and Fig.1(b) show that the agreement between experimental data and the calculated spectrum using the new interaction Sn132ph is reasonable for the studied nuclei. However, it gives an inversion between the ground state and the first excited one of ^{130}Sb. For ^{132}Te and ^{134}I, it gives an inversion between 8^+ and 10^+, and 3^+ and 5^+ states, respectively. The CD-Bonnpn interaction give different energies from the experimental ones.

We have also calculated the electromagnetic properties for ^{130}Sb, ^{132}Te and ^{134}I nuclei, and the results are presented in Tab. 1, Tab. 2 and Tab. 3, in comparison with the available data. The values of electromagnetic properties B(M1), B(E2), τ, μ and Q for the studied nuclei were different in both interactions. Therefore, the calculations of μ and Q moments for ^{130}Sb gave the same values for both interactions. But, for ^{132}Te nucleus, the calculated value of τ was found a very good agreement with experimental one on *Sn132ph* interaction, as shown in table below.

DOI: 10.1051/epjconf/201715401016

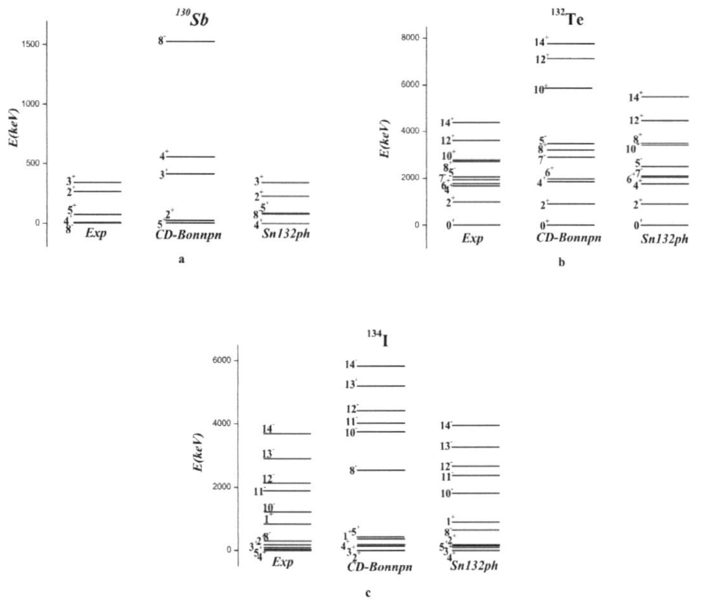

Fig. 1. Calculated energetic spectra using CD-Bonnpn and Sn132ph interactions in comparison with the experimental ones for ^{130}Sb, ^{32}Te and ^{134}I nuclei.

Table 1. Electromagnetic properties for ^{130}Sb nuclei in the vicinity of ^{132}Sn mass region

Ji	*Jf*	*B(M1: μ^2_N)*		*B(E2: e^2fm^4)*		*τ(ps)*		*μ(Ji: μ_N)*		*Q(J_i:efm²)*	
		CD-Bonnpn	**Sn132ph**	**CD-Bonnpn**	**Sn132ph**	**CD-Bonnpn**	**Sn132ph**	**CD-Bonnpn**	**Sn132ph**	**CD-Bonnpn**	**Sn132ph**
3+	2^+	0.28	0.611	38	93.357	3.322	3.306	-1.593	-1.171	25.12	22.60
4^+	2^+	/	/	41.733	83.777	1.469	4.224	-0.6826	-0.4184	30.47	26.41
	3^+	1.445	1.435	37.333	70.108	/	/				
5^+	3^+	/	/	44.927	72.164	25.69	42.22	-0.1419	0.01313	28.65	25.76
	4^+	1.552	1.482	142.772	197.427	/	/				

Table 2. Electromagnetic properties for ^{132}Te nuclei in the vicinity of ^{132}Sn mass region

J_i	J_f	*B(E2: e^2fm^4)*		*τ(ps)*			*μ(Ji: μ_N)*			*Q(Ji:efm²)*	
		CD-Bonnpn	**Sn132ph**	*Exp*	**CD-Bonnpn**	**Sn132ph**	*Exp*	**CD-Bonnpn**	**Sn132ph**	**CD-Bonnpn**	**Sn132ph**
2^+	0^+	218.7	335.1	*2.64*	6.424	3.903	*0.70*	0.6955	6.793	6.793	27.71
4^+	2^+	123.4	395.2	/	5.058	4.119	/	2.487	-27.82	-27.82	15.84
6^+	4^+	197.3	97.49	*209E+3*	340E+3	5.834 E+3	*5.08*	4.930	-65.92	-65.92	-60.23
8^+	6^+	81.36	3.143	/	4.796	29.58	/	-1.076	-45.76	-45.76	20.63
10^+	8^+	0.0006642	42.119	/	7.426	) 0.02958	/	-1.712	71.64	71.64	65.71
12^+	10^+	50.01	78.39	/	8.586	8.313	/	0.9203	33.80	33.80	40.93
14^+	12^+	0.1436	2.933	/	4.806	0.405	/	3.259	-21.51	-21.51	7.531

* Corresponding author: fatima.benrachi@umc.edu.dz

Table 3. Electromagnetic properties for ^{134}I nuclei in the vicinity of ^{132}Sn mass region.

J_i	J_f	*B(+1: μ^2_N)*		*B(E2: e^2fm^4)*		*τ(ps)*			*μ(Ji: μ_N)*		*Q(Ji:efm²)*	
		CD-Bonnpn	**Sn132ph**	**CD-Bonnpn**	**Sn132ph**	*Exp*	**CD-Bonnpn**	**Sn132ph**	**CD-Bonnpn**	**Sn132ph**	**CD-Bonnpn**	**Sn132ph**
2^+	1^+	0.6204	0.002163	149.52	0.11514	*0.04*	/	32.92E+03	2.807	2.276	5.833	-26.76
3^+	1^+	/	/	67.84	152.48	*22.72*	87.02	1333	3.839	2.058	-21.45	-34.43
	2^+	0.2901	2.825E-04	221	3.632							
4^+	2^+	/	/	0.06218	62.222	*3.36*	0.8121E+5	/	3.044	2.992	-19.06	-5.799
	3^+	0.009428	0.03990	28.34	13.618							
5^+	3^+	/	/	7.051	81.44	/	147.4	1535	3.523	3.377	-8.522	0.5724
	4^+	0.02443	0.01132	251.3	215.1							
10^-	8^-	/	/	101.9	115.6	/	0.8585	1.981	4.897	3.472	-6.533	26.62
12^-	10^-	/	/	4.605	5.924	/	0.4785	1.134	7.216	7.09	-18.01	-18.56
	11^-	1.952	1.963	300.5	285.5							
13^-	12^-	0.8026	0.5275	108.8	76.41	/	0.1426	0.4956	7.684	6.529	-25.66	-14.5
14^-	12^-	/	/	41.9	34.72	/	0.1968	0.2329	7.92	8.00	-28.64	-29.05
	13^-	1.074	0.8296	128.4	125.7							

4 Conclusion

In this work, we focused on the study of the mass effect on the nuclear structure of nuclei far from beta stability. More generally, we have explored in a semi-quantitative manner the nuclear force which lead to modification of shell structure in ^{132}Sn region. The properties were derived from two-body effective interactions including monopole effect in the SM framework, by means of Oxbash nuclear structure code using CD-Bonn the original interaction of the code and carrying out some modifications based on the mass factor and monopole effect to get *Sn132ph* new interaction. The resulting energetic spectra using the new interactions are in good agreement with the available experimental data. For the electromagnetic properties, the obtained values using the original and the new interactions are different the three nuclei. But, for ^{132}Te nucleus, the calculated value of τ of the first excited state using *Sn132ph* interaction is in agreement with experimental one.

Authors of this article thanks the organizers of the 3rd International Conference on Theoretical and Experimental studies in nuclear applications and technology 10-12 May 2017 Adana-Turkey for the organization and the support provided during the conference. Special thanks are owed to B. A. Brown for his help in providing us the OXBASH code (Windows Version).

References

1. S. Katcoff, J.A. Miskel and C.W. Stanley, Phys. Rev. **74**, 631 (1948).
2. J. Kathawa, C. Fry, M. Thoennessen, At. Data and Nucl. Data Tables **99**, 22 (2013).
3. S. H. Liu et al., Phys. Rev. C **79**, 067303 (2009).
4. P. Mason et al., Acta Phys. Pol. B **40**, 489 (2009).
5. L. Coraggio et al., Phys. Rev. C **80**, 061303 (2009).
6. N. Laouet, F. Benrachi, Acta Phys. Pol. B **1**, 10 (2017).
7. B. A. Brown et al., Phys. Rev. C **71**, 044317(2005).
8. A. Bisoi et al., *Proceedings of the DAE-BRNS Symp. on Nucl. Phys*. **59**, 104 (2014).
9. S. Das et al., *Proceedings of the DAE-BRNS Symp. on Nucl. Phys*. **60**, 108-109 (2015).
10. E. Hagebo, A. Kjelberg and A. C. Pappas, J. Inorg. Nucl. Chem. **24**, 117 (1962).
11. A. Covello et al., Acta Phys. Pol. B **34**, 2257 (2003).
12. S. Sarkar et al., Phys. Rev. C **81**, 064328 (2010).
13. A. P. Zuker, Phys. Rev. Lett. **90**, 042502 (2003).
14. A. Poves and A. Zuker, Phys. Rep. **70**, 235 (1981).
15. T. Otsuka et al., Phys. Rev. Lett. **105**, 032501 (2010).
16. O. Sorlin, M.G. Porquet, Prog. Part. Nucl. Phys. **61**, 602 (2008).
17. B.A. Brown, *Oxbash for Windows*, MSU-NSCL Report 1289 (2004).
18. R. Machleidt, F. Sammarruca and Y. Song, Phys. Rev. C **53**, R1483 (1996).
19. R. Machleidt, Phys. Rev. C **63**, 024001 (2001).
20. M. Wang et al., "*The AME2012 atomic mass evaluation*," Chinese Phys. C **36**, 1603 (2012).
21. S. I. Sukhoruchkin and Z. N. Soroko, Landolt-Börnstein-Group I *Elementary Particles, Nuclei and Atoms*, H. Schopper (ed) , Springer materials, 2013.
22. V. I. Isakov et al., Phys. of Atom. Nucl, **70**, 818 (2007).
23. K. Heyde, Hyp. Int. **43**, 15 (1988).

EPJ Web of Conferences **154**, 01017 (2017) DOI: 10.1051/epjconf/201715401017
TESNAT 2017

The overview of nuclear energy situation in the World and Turkey

Yusuf Alper Kaplan*, *Merve* Karagöz, *Serhat* Sayılmaz

Osmaniye Korkut Ata University, Energy System Department, Turkey

Abstract The dependence on the energy and its use has increased in every country due to the increasing population and advanced technology. As a result of it, the reserves of fossil fuel have decreased, several energy crises have occurred from time to time and the alternative energy sources have been on the focus. One of these alternative energy sources is nuclear energy. The nuclear power plants, which were built in order to get nuclear energy, have attracted the attention thanks to some disadvantages such as its high cost and emission of radiation while they do not radiate harmful gases towards environment. The nuclear power plants that have already been and are planned to be constructed by a number of countries have become problematic because of the power plant accidents. On one hand, some countries have abandoned the nuclear power plants owing to the accidents mentioned above, on the other hand some other countries have continued to operate the nuclear power plants by claiming the necessity to meet the increasing demand on energy. It is seen that conflicts and problems experienced in the geography in which Turkey is located impacts the energy security of Turkey and it is understood that this situation may have a negative influence on national security of Turkey. Because of all these reasons, actualizing nuclear energy projects are important for Turkey which is dependent in respect of energy.

1 Introduction

Being addicted to gasoline and gas makes the world research energy strategies. Because of the energy crisis countries aim to evaluate energy gender and conflict energy sector, use alternative energy source [1]. In the current context of globalization, the energy used by a country is not anymore, a suitable indicator for measuring the total energy requirements associated with its level of development; the significant variable is the energy consumed worldwide to produce the goods and services demanded by that country, i.e. its energy footprint [2].

Countries started to find nuclear energy solutions due to energy issues [1]. Nuclear energy is a tool which most of the countries can use to get outmaneuvering against of each other. The main reason of this is that nuclear energy has a potential to affect development of countries by providing economic and fiscal returns [3]. Asian countries, EU countries and USA checking new nuclear stations. National development and industries based on energy and new energy sources (nuclear energy) is also important beside traditional and local energy sources in countries national plans and programs [1]. Nuclear energy is very important since it provides about 17% of the electric energy in the world and is used in industry and medical area [4].

Turkey has started discussion about nuclear energy in 1960's. There were some cancelation of processes and negations but arguments are starting again nowadays. Supporting and non-supporting groups are telling their opposing views for Turkey's nuclear energy strategies [1].

2 Nuclear Energy İn The World

The EU Acquis on nuclear energy; The emphasis was on securing safety in nuclear power generation, disabling plants, managing radioactive waste, assessing environmental impact, and addressing technical aspects [5]. The EU has issued directives on the safety of nuclear installations, with particular emphasis on ensuring nuclear safety. As the amount of world energy is at a certain level, developed countries have emphasized the efficient use of energy for a sufficient use of energy. For this reason, developed countries are studying energy efficiency in detail, and they carry out their R & D work with care [6]. In particular, it is important to understand how to generate more power from new generation nuclear power plants, to increase the service life of power plants, Is manifested in studies directed towards. EU countries are working on the same amount of energy to make more research and development work for less waste, greenhouse gas and carbon dioxide emissions with less energy, more warming, less energy-consuming technology systems [7].

There are 449 nuclear reactors in 31 countries shown in Fig. 1. Also, there are 60 power plant is under construction shown in Fig. 2 [3]. USA obtains electricity from %20 nuclear energy and world's most nuclear central country with 99 centrals. This ratio is %77 in France; %19 in Russia with 35 centrals (8 central are

* Corresponding author: yakaplan@osmaniye.edu.tr

DOI: 10.1051/epjconf/201715401017

under construction); %30 in South Korea with 24 centrals (4 central are under construction). China has started 23 nuclear central construction which has 31 central already. There are a lot of nuclear centrals in oil and gas-rich countries even. Such as; USA, Canada, England, Russia for oil-rich and Iran, Russia for gas-rich countries use nuclear energy for electric power production [10].

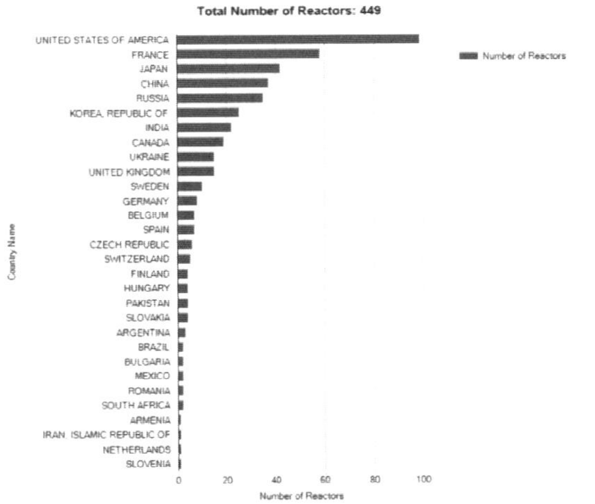

Fig. 1. Nuclear energy reactors for some countries [8]

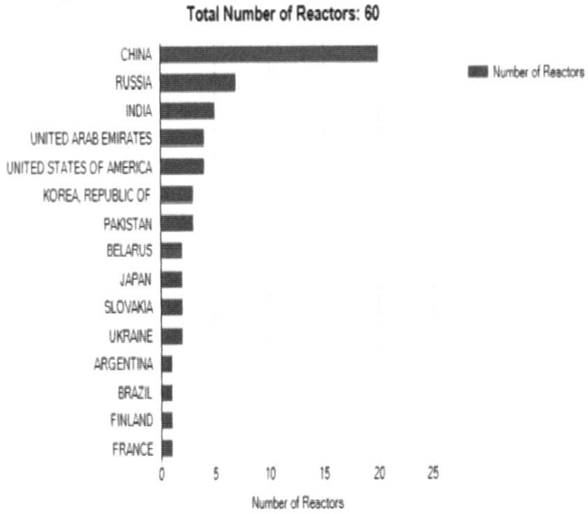

Fig. 2. Under construction reactors [8]

2.1 Nuclear Power in the USA

The USA is the world's largest producer of nuclear power, accounting for more than 30% of worldwide nuclear generation of electricity. The country's 100 nuclear reactors produced 805 billion kWh in 2016, almost 20% of total electrical output. There are four reactors under construction. Following a 30-year period in which few new reactors were built, it is expected that four more new units will come online by 2021, these resulting from 16 license applications made since mid-2007 to build 24 new nuclear reactors Government policy changes since the late 1990s have helped pave the way for significant growth in nuclear capacity [9].

2.2 Nuclear Power in France

France derives about 75% of its electricity from nuclear energy, due to a long-standing policy based on energy security. This share may be reduced to 50% by 2025. France is the world's largest net exporter of electricity due to its very low cost of generation, and gains over €3 billion per year from this. France has been very active in developing nuclear technology. Reactors and especially fuel products and services have been a significant export. About 17% of France's electricity is from recycled nuclear fuel [9].

2.3 Nuclear Power in the Russia

Russia is moving steadily forward with plans for an expanded role of nuclear energy, including development of new reactor technology. An average of one large reactor per year is due to come on line to 2028, balancing retired capacity. Efficiency of nuclear generation in Russia has increased dramatically since the mid-1990s. Exports of nuclear goods and services are a major Russian policy and economic objective. Over 20 nuclear power reactors are confirmed or planned for export construction. Foreign orders totaled $133 billion at the end of 2016 [9].

2.4 Nuclear Power in the China

Mainland China has 36 nuclear power reactors in operation, 21 under construction, and more about to start construction. Additional reactors are planned, including some of the world's most advanced, to give a doubling of nuclear capacity to at least 58 GWe by 2020-21, then up to 150 GWe by 2030, and much more by 2050. The impetus for increasing nuclear power share in China is increasingly due to air pollution from coal-fired plants. China's policy is to have a closed nuclear fuel cycle. China's policy is to 'go global' with exporting nuclear technology including heavy components in the supply chain [9].

3 Nuclear Energy in the Turkey

Turkey, which cannot meet its energy with local energy sources, has turned to nuclear energy as an alternative energy source. It is seen that conflicts and problems experienced in the geography in which Turkey is located impacts the energy security of Turkey and it is understood that this situation may have a negative influence on national security of Turkey. Because of all these reasons, actualizing nuclear energy projects are important for Turkey which is dependent in respect of energy [11]. Fig. 3 shows the plants planned to be installed in Turkey.

Turkey has had plans for establishing nuclear power generation since 1970.Today, plans for nuclear power are a key aspect of the country's aim for economic growth. Recent developments have seen Russia take a leading role in offering to finance and build 4800 MWe of nuclear capacity. Application has been made for construction and operating licenses for the first plant, at Akkuyu, and construction start is expected in 2018.A Franco-Japanese consortium is to build the second nuclear plant, at Sinop. China is in line to build the third plant, with US-derived technology. A small uranium mining project is planned [9].

Fig. 3. The nuclear reactors are planned to be established in Turkey [12]

3.1 Akkuyu NGS

Turkish Electricity Trade & Contract Corporation (TETAS) called for tenders in March 2008, inviting bids for the first nuclear power plant at Akkuyu, near the port of Mersin in Mersin province on the eastern Mediterranean coast. Turkish Atomic Energy Authority (TAEK) issued specifications, allowing for Pressurized Water Reactor (PWR), Boiling Water Reactor (BWR) or Pressurized Heavy Water Reactor (PHWR) types of at least 600 MWe and with 40-year service life. Design certification in country of origin was acceptable, allowing TAEK to concentrate on site-specific aspects of the 4800 MWe project. In the event, only one bid was received from 14 interested parties, this being from Atomstroyexport in conjunction with Inter RAO (both from Russia) and Park Teknik (Turkey), for an AES-2006 power plant with four 1200 MWe reactors [13].

Following commercial advice from TETAS, a government decision was expected in April 2009, but in fact only a series of statements resulted, regarding the cost of power over the first 15 years being too high. Then in August 2009 two agreements between TAEK and Rosatom were signed with much fanfare. One was a nuclear cooperation agreement, the other was a standard one on the early notification on a nuclear accident and the exchange of information on nuclear facilities. These progressed the possibility of a Russian nuclear project at Akkuyu, probably with 25% government equity to dampen the likely electricity price rise. The first reactor was expected to come online in 2016, and others in 2017, 2018 and 2019. However, following a ruling by the country's top legal body, TETAS cancelled the Atomstroyexport agreement and said that a new tender would be launched soon. Instead, the parties proceeded to a direct high-level agreement.

Atomstroyexport is general contractor for construction. Early in 2017 the energy minister suggested 2017 start. In June 2016 Rosatom said that construction would start in 2018. In August 2016, the government said that it would classify the project as a strategic investment, allowing favorable tax treatment.

TAEK on 9 February 2017; Akkuyu approved the field parameters to be used in the design of the plant, which has been determined by Nuclear Inc. in detail and which has been justified and submitted to the approval of the Agency within the scope of the Akkuyu Nuclear Power Plant Field Parameters Report (SPR). The report prepared in line with the results of the engineering studies obtained on the nuclear power plant site to be established, contains the precise values acquired for reflection in Akkuyu NGS design Project [14].

3.2 Sinop NGS

Since February 2008 preparatory work has been under way at Sinop on the Black Sea coast to build a second nuclear plant there, along with a €1.7 billion nuclear technology centre. A 5000-5600 MWe nuclear plant there is expected to cost about $22-25 billion.

A feasibility study is due for completion in mid-2017, and the announced timeline is for construction start in 2017 and operation from 2023. These are likely to be the first Atmea1 units built. They would be designed for load-following and use the same steam generators as Areva's large EPR (but three instead of four). Cost is projected at $4.89 billion each. MHI has set up a new business unit, the Turkey Nuclear IPP Development Department, to handle the feasibility study, negotiate contracts, prepare financing and undertake other functions in Turkey. In April 2014, an intergovernmental agreement was signed with Japan for development of the plant. MHI will lead the engineering-procurement-construction (EPC) consortium [9].

3.3 Igneada

There are plans to build further nuclear capacity at another site, as part of 100 GWe required by 2030. TAEK has identified Igneada in Kirklareli province on the Black Sea, 12 km from the Bulgarian border, and this was confirmed in October 2015. Akcakoca between it and Sinop was also considered. When agreement for the development of the Sinop plant was to have been finalized, the energy ministry was to announce the site for the third plant with an invitation for expressions of interest to be issued by the end of 2013. This didn't happen and in October 2014 the prime minister said that the project would be substantially indigenous, with construction start about 2019 [9].

4 Expectations of energy future of Turkey

Integration of nuclear energy into our energy supply sources will be one of the major means for decreasing the risks emerging from dependency to imported fuels by satisfying the increasing electric energy demand. MENR targets that the share of nuclear power plants in electric energy generation will reach at least 10% by the end of 2023. To achieve this target, it is planned that two nuclear power plants will be in operation and one will be under construction by the end of 2023.

According to the Agreement with the Japan, 4 units of ATMEA-1 type reactor, each with 1,120 MW power, will be constructed in Sinop province. Total installed capacity of Sinop NPP is 4480 MWe and total lifetime of each units is 60 years. It is expected that first two units of Sinop NPP will be in operation in 2023 and 2024 and other two units in 2027 and 2028. For third nuclear power plant, site selection studies are ongoing [15].

Tablo 1. Planned Nuclear Power Plants

Station/Project Name	Type	Capacity (MWe)	
Akkuyu NPP-1	VVER-1200	1200	2016-2020
Akkuyu NPP-2	VVER-1200	1200	2017-2021
Akkuyu NPP-3	VVER-1200	1200	2018-2022
Akkuyu NPP-4	VVER-1200	1200	2019-2023
Sinop NPP-1	ATMEA-1	1120	2019-2023
Sinop NPP-2	ATMEA-2	1120	2020-2024
Sinop NPP-3	ATMEA-3	1120	2023-2027
Sinop NPP-4	ATMEA-4	1120	2024-2028

5 Conclusion

In Turkey and in the world, the current state of nuclear energy is generally evaluated using new data. The energy resources of the world are rapidly depleting, problems such as costs and global warming are increasing. In this context, nuclear energy studies should be supported, but studies should be carried out with great care to minimize accidents. It is an inevitable fact to put nuclear energy into operation so that Turkey can provide energy independence and energy supply security. Public awareness of nuclear waste should be raised. The wrong judgments on the nuclear issue must be resolved. Together with these developments, it is necessary to exploit alternative sources in Turkey and optimally utilize these resources.

Acknowledgements

The study is supported by Osmaniye Korkut Ata University through the projects OKUBAP-2016-PT3-001.

References

1. Yıldırım M.ve Örnek İ.; Gaziantep Üniversitesi Sosyal Bilimler Dergisi 6(1):32-44 (2007)
2. Arto, I., Capellán-Pérez, I., Lago, R., Bueno, G., & Bermejo, R. (2016). Energy for Sustainable Development, 33, pp: 1-13.
3. Bahçe, A.B., and Gümüş. Usak University Journal of Social Sciences, 9(2), 2016
4. Yıldırım, M., & Ornek, I. (2007). Gaziantep University Social Sciences,6(1), pp :32-44.
5. Council Regulation (EURATOM), 1999, No 2587/1999, (Article 41 of the Treaty establishing the European Atomic Energy Community, December 1999), s.2.
6. M. Toman, B. Jemelkova, Resources for the Future, Vol.41, (April 2003), s.1.
7. B. Colijn and B. Van Ark, "Growth in Six European Countries", Neujobs Deliverable, Vol.3 No.3. (2014), s.6.
8. https://www.iaea.org/PRIS/WorldStatistics/OperationalReactorsByCountry.aspx
9. World Nuclear Assocation Nuclear Power http://www.world-nuclear.org/
10. http://www.enerji.gov.tr/
11. Furuncu, Y. (2016). Cumhuriyet Science Journal, 37, pp: 207-216.
12. Kaplan, Y.A., and C. Aladağ. *EPJ Web of Conferences*. Vol. 128. EDP Sciences, 2016.
13. Yılmaz, E.A. Republican University Literature Faculty Social Sciences Journal , 39.1 (2015), pp:227-245.
14. http://www.akkunpp.com/
15. http://www-pub.iaea.org/

 DOI: 10.1051/epjconf/201715401018

A review of TLD's zero-count based on temperature and radiation history of them

Nina TUNÇEL[1,2] *, *İsmail KARAKUŞ*[1], *Ertuğrul DÜNDAR*[1], *Özlem TOYKAN ÇİFLİKÇİ*[1]

[1]*Akdeniz University, School of Medicine, Radiation Oncology Department, Antalya/ Turkey*

[2]*Akdeniz University, Science Faculty, Physics Department, Antalya/ Turkey*

Abstract. In order to review the background value of a group of TLD-100, the zero-count values were collected from the first reading and after seven years after using them in experimental irradiations. The zero-counts for these two conditions were collected, as well as the dependence of temperature and radiation history of this group was evaluated. This study recommended that after frequently irradiation of TLDs the rearrangement for obtaining zero-count value must be performed before using TLDs in a new irradiation examination. Regarding on memory of radiation and thermal history, the sensitivity can change after receiving a large dose of radiation and undergoing readout. Additional annealing would be necessary to restore the original sensitivity. TLDs must be used under reproducible conditions to obtain consistent results.

1 Introduction

Thermoluminescence (TL) or thermally stimulated luminescence (TSL) is the light emission after removal of the source of exciting energy light, x-rays, or other radiation; the free electrons may be trapped at an energy level higher than their ground state by application of thermal energy [1,2]. The transition of electrons directly from a metastable state to ground state is forbidden. The metastable state represents a shallow electron trap and electrons returning from it to the excited state require energy. This energy can be supplied in the form of heat (thermal stimulation). The probability per unit time that a trapped electron will escape from a metastable state to an excited state is governed by the Boltzmann equation [3,4]. First, the intensity of thermoluminescent emission does not remain constant at constant temperature, but decreases with time and eventually ceases altogether. Second, the spectrum of the thermoluminescence is highly dependent on the composition of the material and is only slightly affected by the temperature of heating [5]. In the usual thermoluminescence experiments, the system is treated at a temperature at which the phosphorescence intensity is low, and later heated through a temperature range where the phosphorescence intensity is bright, until a temperature level at which all the charges have been thermally excited out of their metastable levels and the luminescence completely disappear [6,7]. The thermoluminescence emission mainly is used in solid state dosimetry for measurement of ionizing radiation dose. In order to stabilize the trap structure, annealing processes based on temperature and time pattern are used to produce the lowest intrinsic background and to obtain the highest efficiency. Before using a new thermoluminescent (TL) crystal material for dosimetric proposes, it is needed to perform a thermal treatment process, this thermal treatment consist in many steps as initialization treatment, annealing treatment and post-irradiation treatment [5, 8-10]. This thermal treatment is used for new thermoluminescent materials which have not been used. The aim of this thermal treatment is to stabilize the trap levels, so that during subsequent uses the intrinsic background and the sensitivity are both reproducible. The time and temperature of the initialization annealing are, in general, the same as those of the standard annealing. Standard annealing process is used to erase any previous residual irradiation effect which is supposed to remain stored in the crystal after the readout. It is carried out before using the TLDs in new measurements. The aim of this thermal treatment is to bring back the traps-recombination centres structure to the former one obtained after this procedure. At the end of annealing procedure, the dosimeters are read to check the background signal or zero-count. The background depends on the high voltage that applied to the photomultiplier tube, on the temperature stability of TL reader and also reuse of TL material [1]. The background values determined for each dosimeter have to be collected, which is memorized in a file, so that they can be used for the successive tests [3,11,12]. In many cases an average background value is considered for the whole batch and then subtracted from each individual reading of the irradiated TLDs. This procedure is valid when the background is very low and constant for the whole batch. Post-irradiation

* Corresponding author: ninatuncel@akdeniz.edu.tr

treatment is used to erase the low temperature peaks, if they are found in the glow curve structure. Such low temperature peaks are normally subjected to a quick thermal fading and many times this value is not been included in the readout to avoid any errors in the dose determination [9,13].

2 Aim

In order to review the background value of a group of TLD-100, the zero-count values were collected from the first reading and after seven years after using them in experimental irradiations. Based on group average and standard deviation of reading counts, the zero-counts for these two conditions were calculated as well as the dependence of temperature and radiation history of this group was evaluated.

3 Materials and Method

A hundred TLD-100 type were selected for first applications on 2010. Lithium fluoride doped with magnesium and titanium (LiF:Mg,Ti) is manufactured in the form of solid pellets of 4.5 mm diameter, 0.9 mm thickness under the code name MTS (MTS-N Poland) was annealed in a heat- time programmed oven. The TLDs were annealed at 400°C for 1h and 100°C for 1h using a PTW high temperature oven (PTW Freiburg GmbH) for initial temperature treatment. And before reading TLDs the pre-read heating processed was performed (100°C for 1h). The luminescent signals for zero-counts were counted by an automated TLD reader RADOS RE-2000RT (RadRro Int. GmbH Germany). The reading arrangement was based on 4 sub-groups. So, the zero-count value was obtained for these sub-groups. After seven years irradiation experiments the amount of TLDs was 80 so the collection of zero-count values was done for these remnants. This process was repeated three times for obtaining consistency of zero-count values. The average, standard deviation (SD) and percent error of these values were calculated.

4 Results

The zero-count values for 4 sub group of TLDs on 2010 were shown in Table 1. The general use of TLDs for routine dosimetry purposes was quite straightforward. Simplification and convenience was often implemented that make operations easier for the users but may not be consistent with the optimization of TLD performance and stability because of the more complex materials in terms of its characteristics especially with regard to the electron-hole trapping centres. The various trapping centres are defined by different energy depths (or temperature), but they are not all thermally independent, so that there may be transitions among traps such that one energy trap may deplete while another is enhanced, depending on thermal effects. One of the most important influencing factors are the annealing history to which the dosimeters have been subject. The recommended full anneal process from the high temperature (400° C) and cool-down period to the lower temperature anneal for LiF:Mg,Ti TLD-100 type result in sever change on TLD sensitivity. In this study, it is important to be consistent and use the same method every time that the TLDs were annealed. So, the reproducibility regarding annealing process was constant. Even with care in the annealing process, these TLDs were used in radiotherapy experiments many times that irradiated on low and high energy photons along seven years. After this period, for quite objective analyzing on TLD sensitivity the zero-count values were collected three times consecutively Table 2.

Table 1. The Zero-Count values at initial use on 2010.

Sub Group	**The Zero-Count on 2010**			
	1.	**2.**	**3.**	**4.**
Average	3539	3339	3389	4165
SD	541	113	143	348
% error	15	3	4	8

Table 2. The Zero-Count values after seven years on 2017.

Sub Group	**First reading/the Zero-Count on 2017**			
	1.	**2.**	**3.**	**4.**
Average	9019	9069	9519	12362
SD	1533	2348	1883	3022
% error	17	26	20	24
Sub Group	**Second reading/the Zero-Count on 2017**			
	1.	**2.**	**3.**	**4.**
Average	6280	6494	6588	6805
SD	1783	1429	1919	1635
% error	28	22	29	24
Sub Group	**Third Reading/the Zero-Count on 2017**			
	1.	**2.**	**3.**	**4.**
Average	3338	3477	3673	4075
SD	1169	973	1300	1233
% error	35	28	35	30

5 Conclusions

This study recommended that after frequently irradiation of TLDs the rearrangement for obtaining zero-count value must be performed before using TLDs in a new irradiation examination. Regarding on memory of radiation and thermal history, the sensitivity can change after receiving a large dose of radiation and undergoing readout. Additional annealing would be necessary to restore the original sensitivity. TLDs must be used under reproducible conditions to obtain consistent results.

References

1. S.W.S. McKeever, *Thermoluminescence of Solids*. Cambridge University Press, Cambridge, United Kingdom, 1985.
2. G.F. Knoll, *Thermoluminescent Dosimeters*. Section G, Chapter 19, in *Radiation Detection and Measurement*, Third Edition. John Wiley & Sons, New York, 1999.

DOI: 10.1051/epjconf/201715401018

3. Y.S. Horowitz, *Thermoluminescence and Thermoluminescent Dosimetry*. (ed). CRC Press, Boca Raton, Florida. 1984.
4. F.H. Attix, *Introduction to Radiological Physics and Radiation Dosimetry*. Wiley-VCH Verlag GmbH & Company, 2004.
5. S.W.S. McKeever, M. Moscovitch, and P.D. Townsend, *Thermoluminescence dosimetry materials: properties and uses*. Nuclear Technology Publishing, Kent, England, 1995
6. R. Chen, and S.W.S. McKeever, *Theory of Thermoluminescence and Related Phenomena*. World Scientific Publications Singapore, 1997.
7. Y. S. Horowitz, Phys. Med.Biol. **26**(5), 765-824 (1981).
8. J.R. Cameron et al., *Thermoluminescent Dosimetry*. University of Wisconsin Press, Madison, Wisconsin, 1968.
9. C. Furetta and P.S. Weng. *Operational Thermoluminescence Dosimetry*. (World Scientific Publishing, Hackensack, New Jersey, 1998).
10. P.W. Frame, Health Phys. 87(2), 111–135 (2004).
11. C.A. Carlsson, Phys. Med. Biol. **14**(1), 107-118 (1969).
12. B.K.A. Martensson, Phys. Med. Biol. **14**(1), 119-130 (1969).
13. A.P. Turner and D.W. Anderson, Phys. Med. Biol. **18**(1), 46-52 (1973).

Electron beam irradiation processing for industrial and medical applications

Zehra Nur Ozer*

Afyon Kocatepe University, Physics Dept., Afyonkarahisar, Turkey

Abstract. In recent years, electron beam processing has been widely used for medical and industrial applications. Electron beam accelerators are reliable and durable equipments that can produce ionizing radiation when it is needed for a particular commercial use. On the industrial scale, accelerators are used to generate electrons in between 0.1-100 MeV energy range. These accelerators are used mainly in plastics, automotive, wire and electric cables, semiconductors, health care, aerospace and environmental industries, as well as numerous researches. This study presents the current applications of electron beam processing in medicine and industry. Also planned study of a design for such a system in the energy range of 200-300 keV is introduced.

1 Introduction

Irradiation process has been used in industry for many years. Electron beam irradiation is very similar to gamma radiation in terms of ionizing energy but is different in terms of dosage rates and penetration. By comparison to as gamma rays, electron beam processing has several more advantages (Table 1). Specially, electron beam technology is referred as safe and cheap technology.

Table 1. Electron beam advantages compared to gamma rays [1].

Feature	Benefit
Faster revisal	Reduced inventory
Flexibility	Quick dose changes
Less material degradation	Reduced risk of product damage
Uniform, controlled dose	Precise dose delivery with no shielding from other products
Non radioactive dose	No security, transportation / on-off technology
Re-sourcing is not required	No radioactive cobalt resupply downtime

Electron beam irradiation is a process for treating materials with energetic electrons that are produced by an accelerator. The electron beam accelerators cover a wide range of beam currents and energies to produce a large scale of products. Using electron beam processing, the chemical and physical properties of material is modified and the product quality is enhanced.

Energy utilization of electron beam processing is rather higher. Catalysis is not required in this system because energy is directly injected in to the material. Electron beam has directivity and high processing performance. The ratio of energy given to the material per a unit length is higher than that of gamma rays. In addition to that, the absorbed dose rate is very large since the electron beam has directivity to the direction of acceleration. Moreover, reaction control, operation and maintenance are easy for electron beam processing system. The reaction can be easily started or stopped by turning on and off the switch and by setting several parameters the system can be operated easily [2-4].

Electron beams are commonly used in industry for medical, environmental and material processing applications. In medical industry, electron beam processing system is used for medical equipment sterilization, water treatment, food preserving and pollution prevention for environmental industry [4-7]. In material processing industry, the major application of electron beam is grafting, curing, degrading and cross linking. According to the studies, 55% of the market of industrial electron beam accelerator comes from cross linking process that contains wire and cable tubing, shrink film and tires [7-10]. Here, we present a short review on the literature of industrial and medical electron beam accelerators and the important parameters in applications.

nergy and current are the two main characteristics of electron beam accelerators in industrial applications. 0.15 to 10 MeV energy ranges of electron beams are used for processing of various applications. Low energy electron accelerators having energies 0.15 to 0.5 MeV are generally used for treatment of surfaces and irradiation of coatings and polymeric films. Medium energetic ones produce electrons that have energies of 0.5-5 MeV. High energetic ones have energies between 5-10 MeV and they are used for sterilization of medical products and processing treatment of food.

* Corresponding author: zehraerengil@aku.edu.tr

2 Electron Beam Processing System

Low energy accelerators are known as non-scanning type of electron beam processing. The source is linear source so no scanning is necessary. It is like a wide shower of electron beam to cover necessary area of the product. The medium accelerators are commonly based on transformer type and acceleration tube to produce scanned beams. In processing industry for high energy electron beam accelerators; microwave linear or radio frequency devices are used [8,10].

The principal of acceleration is the basic system to produce an electron beam and accelerated them in the electron accelerator. Tungsten filament is heated at the negative cathode and electron is then emitted as a free electron called by thermionic emission. The free electron flow is accelerated by electric field and gain energy in electron volt. The accelerated electrons escape through the window and proceed towards the target material. Electrons are accelerated and given energy to be irradiated to target material by electron beam processing system. The processing system is consisting of power supply unit, accelerator tube, scanning system, irradiation window, vacuum system, product handling system and safety system.

There are different types of accelerators available with different system designs but they all use high voltage to accelerate electron in vacuum. Power supply unit is the most important part to generate high voltage of direct current. High voltage generator designed to follow the specification of the accelerator commonly based on the transformer type. The electron accelerator is usually located inside the high voltage vacuum. The main function of accelerator tube is to generate electrons and accelerate them to gain energy in electron volt. The accelerated electrons escape through the window and proceed towards the target material.

The scanning system contains focusing lens, deflection system, valve, scanning chamber, irradiation window. The beam is controlled by the focusing lenses and it is deflected through a specific angle according to design of the machine. The electron beam is allowed to be scanned sensitively for required width at target product area. Then the electron beam is applied to the material by the irradiation window. The window should be thin enough to allow electron beam to penetrate easily and reduce the heat produced by the lost electron energy while passing the window.

Vacuum system is important to prevent accelerated electron from collide with air molecules and also avoid filament to damage. The role of product handling system is to set and monitoring the quality of product. It also can shoe the status of the operation or critical process condition when there are any problems inside the machine. When high energy electron hits to the target, the x-ray radiations are possible to generate. At this point shielding is important for accelerators. Most of the accelerators are made out of lead or steel for this purpose.

3 Important variables for electron beam processing

Variables are important to identify the control of the process in order to predict the production and quality of the product. These variables are:

- Absorbed dose
- Electron beam energy
- Depth of penetration
- Line speed

Absorbed dose is amount of energy that a material absorbs from processing deposited into a specific mass of material. The unit of absorbed does is in gray (Gy) usually. Absorbed dose is proportional to the ionizing energy delivered per unit mass of material. Absorbed does requirements for various industrial processes cover a wide range (0.1 kGy to 1000 kGy). 10 kGy to 150 kGy is needed for most of the applications including sterilization, polymerization, grafting, cross linking whereas less than 3 kGy is needed to control of fungi, bacteria or parasite [5,8,10].

Electron beam energy plays important role on the process. It determines how deep the energy will penetrate into the material. The penetration in the material increases linearly with the incident energy. Materials with higher electrons per unit masses have higher absorbed doses near the entrance surface but lower electron ranges. Penetration into materials of different density can be estimated by multiplying the penetration depth, found from the curves, by the ratio of the density of water to the density of the material. These curves show the penetration for different accelerating voltage to the depth of penetration in a material with mass density equal to that of water [12-14].

Line speed shows the processing rate of the material under the beam. This parameter increases with the increased beam current and decreases with the beam width and absorbed dose.

The appropriate type of industrial application depends on the electron energy and beam power requirements of the process. Electron accelerator types used for the industrial applications are high voltage direct current, microwave and radio frequency systems [10-14].

4 Results

Electron accelerators of several types have been developed and are being used for industrial and medical applications. Currently, there are over thousand industrial electron accelerators in operation in the world wide. These are being widely used for applications such as polymer processing, surface curing, sterilization of medical products and food irradiation.

The low energy electron accelerators can be used in coating machines, printing presses, laminating machines and so on. Medium energy electron accelerators are used to process thick plastic and rubber sheets, wire and cable insulation, plastic tubes and fiber-reinforced composites. Cross linking of wire and cable insulation is the main application. The higher energetic ones (up to 10 MeV) are usually used for sterilization, food irradiation and fine tuning of semiconductors [11-12].

Operation parameters; electron energy, beam power, irradiation width and dose rate can be precisely matched to the demands of the industrial process.

The aim of this study to give an overview of the industrial electron beam apparatuses and applications. This review on the application of electron beams in industry will help to identify potential areas in further research. We have planned to design and simulate a low energy electron curtain accelerator in low energy regime of 100-300 keV for processing the materials surface and realizing processes in gases. Further work on the design and simulation of electron accelerator is in progress.

References

1. http://ebeamservices.com/resources/advantages-of-electron-beam-over-gamma/
2. R. Mehnert, Nuclear Instruments and Methods in Physics Research B **113**, (1996).
3. M. Silindir, A.Y. Ozer., FABAD J Pharm Sci **34.34** (2009).
4. N.A. Abdul Manaf, Project, University of Liverpool, Phys 379.
5. H. Lung et al., Trends in Food Science and Technology, (2015).
6. Accelerators for America's future, US Department of Energy, (2010).
7. M.R. Cleland, R.A. Galloway, New York, IBA Industrial Inc. (2009).
8. M. Kashiwagi, Y. Hoshi, Japan, NHV Corporation, Vol. 75, (2012).
9. B. Laurell, E. Föll, RadTech Europe, Switzerland, (2011).
10. M.R. Cleland, NY, USA: IBA Technology Group, (2005).
11. R.W. Hamm, M.E. Hamm, World Scientific Publishing, (2012).
12. S. Sabharwal, Proceedings of PAC2013, Pasadena, CA USA, (2013).
13. A.J. Berejka, M.R. Cleland, M. Walo, Rad. Phys. And Chem. AJB 7, (2013).
14. B. Bogdanovitch, V. Senioukov, A. Koroliov, K. Simonov, Proceedings of the 1999 Particle Accelerator Conference, NY, (1999).

Reaction cross section calculation of some alkaline earth elements

Eyyup Tel[1,*], *Yusuf* Kavun[2], *Ismail Hakki* Sarpün[3]

[1]Osmaniye Korkut Ata University, Department of Physics, Osmaniye, Turkey
[2]Manisa Celal Bayar University, Department of Physics, Manisa, Turkey
[3]Afyon Kocatepe University, Physics Dept., Afyonkarahisar, Turkey

Abstract: Reaction cross section knowledge is crucial to application nuclear physics such as medical imaging, radiation shielding and material evaluations. Nuclear reaction codes can be used if the experimental data are unavailable or are improbably to be produced because of the experimental trouble. In this study, there action cross sections of some target alkaline earth elements have been calculated by using pre-equilibrium and equilibrium nuclear reaction models for nucleon induced reactions. While these calculations, the Hybrid Model, the Geometry Dependent Hybrid Model, the Full Exciton Model, the Cascade Exciton Model for pre-equilibrium reactions and the Weisskopf-Ewing Model for equilibrium reactions have been used. The calculated cross sections have been discussed and compared with the experimental data taken from Experimental Nuclear Reaction Data library.

1 Introduction

Beryllium (Be), magnesium (Mg), calcium (Ca), strontium (Sr), barium (Ba) and radium (Ra) are alkali earth metal that is located at 2A group of periodic table. These elements are usually shinny and silver in their pure forms are rarely pure. Alkali metals are used in many areas and the use of these metals is increasing as a result of research. These metals are used in nuclear reactors, computer parts, lighting tubes, and rocket fuel, military purposes, our life cycle, health, strength of bones and teeth and plays fundamental role on our body.

In this study, we aim to investigate equilibrium and pre-equilibrium effects of some alkaline earth elements for nucleon induced reactions. For this purpose, ALICE/ASH [1], CEM 95 [2] and PCROSS [3] computer codes are used. The calculated cross sections have been discussed [10] and compared with the experimental data takenfrom Experimental Nuclear Reaction Data library [4].

2 Material and Methods

In this paper, we have calculated reaction cross section of several alkaline earth elements. While these calculations, the Hybrid Model [5], the Geometry Dependent Hybrid (GDH) Model [6], the Full Exciton Model [7], the Cascade Exciton Model [8] for pre-equilibrium reactions and the Weisskopf-Ewing (WE) Model [9] for equilibrium reactions have been used.

By neglecting angular momentum according to WE model equilibrium emission is calculated. Binding energies, inverse reaction cross-section, the pairing, and the level density parameters are the basic parameters of this model. According to the exciton model, after the first interaction between the incoming particle and the target nucleus, the excited system can reach the balance after a series of successive steps of increasing complexity, it may be possible to publish each of these steps. Hybrid and GDH models were modified using Pauli and surface correctionsthat were based on exciton model initially. The Cascade Exciton Model (CEM) accepts that reactions occur in three phases. The first phase is a transition in nuclear levels. The second phase corresponds to the pre-equilibrium state, and the third phase corresponds to the equilibrium state. These three components contribute to the experimentally measured values. Detailed description of this model can be found in their references.

3 Results and Discussion

Comparison of calculated cross sections ^{44}Ca(p,n) ^{44}Sc reaction with EXFOR are given in Fig. 1. PCROSS code calculations do not good fit experimental values after 12 MeV energy. CEM 95 model results are good harmony with Exfor data after 20 MeV energy region. WE Model have same geometry with experimental data up to 20 MeV. GDH and Hybrid Model calculations are best options for this reaction.

As can be seen in Fig. 2., all calculated results have a same geometry with EXFOR values, however, CEM follow experimental data from below. Hybrid and GDH model calculations are best options for these reactions.

DOI: 10.1051/epjconf/201715401020

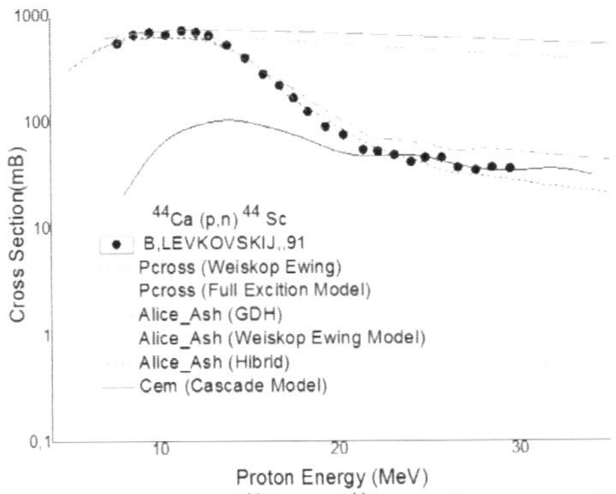

Fig. 1 Comparison of ^{44}Ca(p,n) ^{44}Sc reaction cross section calculations with EXFOR

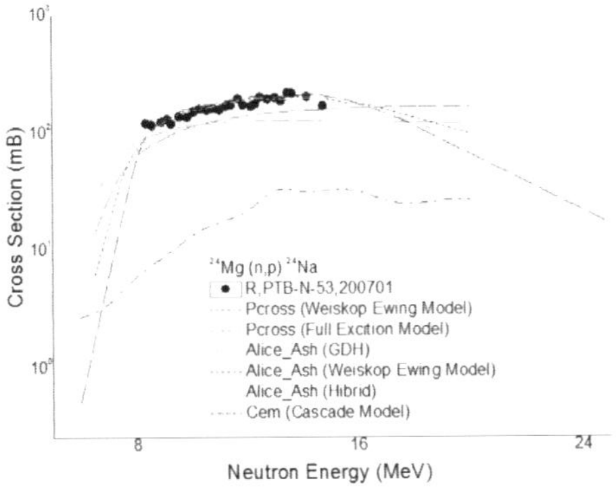

Fig. 2. Comparison of ^{24}Mg(n,p)^{24}Na reaction cross section calculations with EXFOR

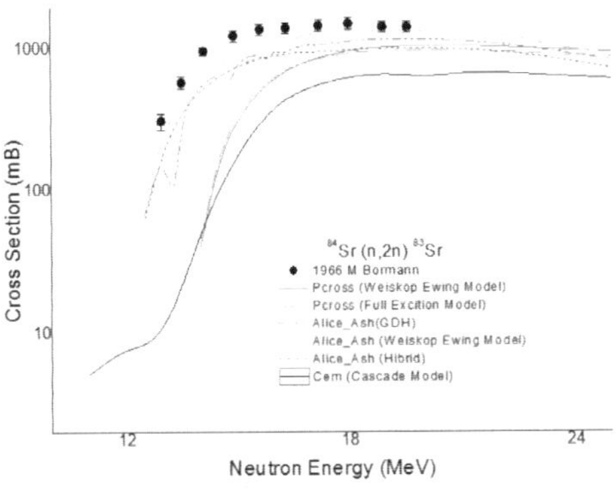

Fig. 3. Comparison of ^{84}Sr(n,2n) ^{83}Sr reaction cross section calculations with EXFOR

Comparison of ^{84}Sr(n,2n)^{83}Sr reaction cross section calculations with EXFOR are given in Fig. 3. All model calculations have a same shape with experimental data but Cascade Exciton Model and PCROSS code calculations follow from below. ALICE/ASH results gave close results with each other. They are good harmony with EXFOR values. ALICE/ASH WE Model are the best option for this reaction.

As can be seen in Fig. 4. ALICE/ASH WEModel and PCROSS code calculations are good harmony with experimental data up to 18 MeV energy. CEM 95 code results have a good harmony experimental data but follow them from below up to 23 MeV energy. ALICE/ASH pre-equilibrium calculations fit experimental data perfectly.

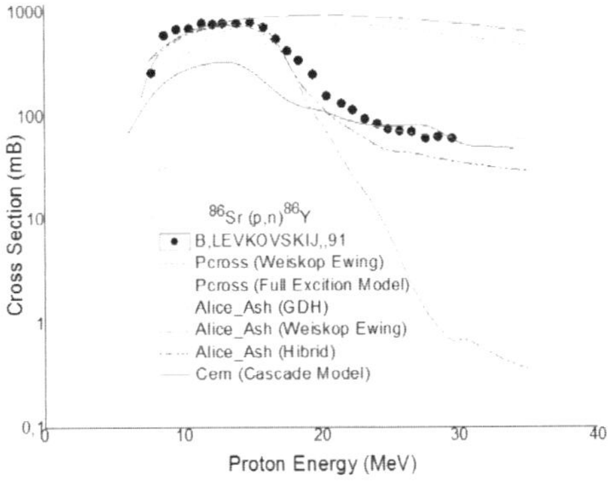

Fig. 4. Comparison of ^{86}Sr(p,n)^{86}Y reaction cross section calculations with EXFOR

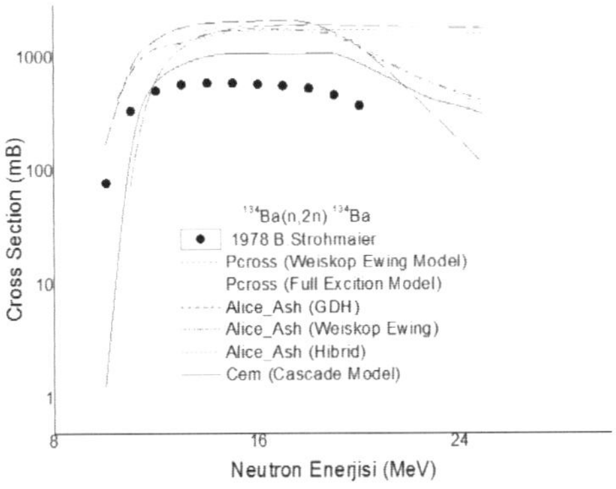

Fig. 5. Comparison of ^{134}Ba(n,2n)^{134}Ba reaction cross section calculations with EXFOR

Comparison of ^{134}Ba(n,2n)^{134}Ba reaction cross section calculations with EXFOR are given in Fig. 5. All model calculations are good harmony with experimental data but follow EXFOR values from above. As can be summarized;

1. Generally PCROSS code calculations are harmony with EXFOR values up to 18 MeV nucleon energy.
2. Cascade Exciton Model calculations have a same geometry shape with experimental databut follow them from below.
3. ALICE/ASH WE Model calculations are in good harmony with EXFOR values up to 18-20 MeV energy region.
4. Pre-equilibrium calculations from ALICE/ASH code can be chosen if the experimental data are unavailable.

Acknowledgment

The study was supported by Osmaniye Korkut Ata University (OKU) Science Research Projects Coordination Unit with the grant number: OKUBAP-2017-PT3-022.

References

1. Broeders, C.H.M., Konobeyev, A. Yu., Korovin, Yu. A., Lunev, V.P., Blann, M., For schungszentrum Karlsruhe in der Helmholtz-Gemeinschaft Wissenschaftliche Berichte FZKA 7183 (2006)
2. Mashnik, S. G. CODE CEM95. Bogoliubov Laboratory of Theoritical Physics, Joint Institute for nuclear research. Dubna, Moskow ,141980, Russia
3. Capote, R., Osorio, V., Lopez, R., Herrera, E., and Piris, M., Final Report on Research Contract 5472/RB, INDC (CUB)-004, Distr,L, (1991)
4. Brook haven National Laboratory, National Nuclear Data Center, EXFOR/CSISRS (Experimental Nuclear Reaction Data File). Database Version of October 11, 2017, (http://www.nndc.bnl.gov/exfor/)
5. Blann, M., Phys.Rev. Lett. 27, 337.
6. Griffin, J.J., Phys. Rev. Lett. 17, 478. (1971)
7. Ribansky, I., Oblozinsky, P., Betak, E., Nucl. Phys. A 205, 545.(1973)
8. Gudima, K.K., Mashnik, S.G., Toneev, V.D., JINR Communication P2-80-774. Dubna; Cascade-Exciton Model of Nuclear Reactions: Comparison of Experiment. P2-80-777, Dubna. (1980)
9. V.F. Weisskopf and D.H. Ewing, Phys. Rev. 57 472 (1940)
10. Özdoğan, H., Çapalı, V., Kaplan, A., J. Fusion Energ. 34 (2), 379-385 (2015)

An Ab-initio study on UAl for nuclear applications

Yasemin Ö. Ciftci*, *Aynur* Özcan, *İrem Ö.* Alp

Gazi University, Department of Physics, Ankara, TURKEY

Abstract: The structural, elastic, electronic and bonding nature and vibrational properties of UAl in B2 structure under zero pressure have been investigated by performing first principles calculations using density functional theory. The exchange-correlation potentials were treated within the generalized gradient approximation (GGA). The calculated quantities are agreed well with the available results. The electronic properties such as band structure and density of states reveal that UAl is metallic in nature with large overlap at Fermi level. The single-crystal elastic stiffness constants of UAl are investigated using stress-strain method. Present results for elastic constants show that this compound is mechanically metastable which is agreement with previous study [3]. The chemical bonding is interpreted by calculating the density of states and electron density distribution analysis. UAl has metallic bonding characteristic. The presented phonon dispersion curves and one-phonon DOS is also confirmed that this compound is dynamically unstable.

1. Introduction

Dispersion nuclear fuels are being developed as fuels for research reactors, which are consists of U–Mo particles dispersed in an Al matrix. The product of the reaction UAlx system has become the subject of numerous studies since the middle of last century owning to its outstanding role in nuclear fuel development [1,2]. Therefore, in order to fully exploit the product material, it is imperative to obtain a good understanding of the physical properties of the UAlx system, and in particular, their structures and dynamic properties. A detailed theoretical study of structural, electronic, and elastic properties of cubic UAlx (x=1,2,3) have been presented employing the pseudopotential plane-wave method based on density-functional theory [3].

In this work, we have been discussed the structural, elastic, electronic and vibrational properties of the uranium-aluminium intermetallic UAl in CsCl crystal structure by analysing first principles calculations. In this study, lattice parameter is quite convenient with literature and the electronic band structures are determined with density of states and this compound exhibits metallic in the nature. Additionally, the structure is dynamically stable considerable as vibrational part of the study.

2. Calculation method

All results have been reported here calculated using Vienna Ab-initio Simulation Package (VASP) which implements Density Functional Theory (DFT) [4-6]. The plane wave based pseudo-potential method has been used for the total energy calculations. The electron-ion interaction and the exchange correlation energy are described under the Projector Augmented Wave (PAW) method and the generalized gradient approximation (GGA) of Perdew Burke Ernzerhof (PBE) scheme, respectively [7-9]. All-electronic projector augmented wave potentials are employed for the elements U and Al. The cut off energy for the plane wave basis set is fixed at 600 eV. For all the structures both the lattice parameters and the atomic coordinates have been relaxed ensuring that energies converge to a precision of less than 10^{-10} eV/atom and Hellmann–Feynman forces to less than 10^{-6} eV/Å. The k-point meshes are constructed using the Monkhorst-Pack scheme and the 14x14x14 k-point mesh is used for the primitive cell for Brillouin zone sampling [10].

3. Results and discussions

3.1 Structural and elastic properties

The crystal structure of UAl alloy is shown in Fig. 1 in which the Uranium atoms are grey colour and aluminium atom is red one in the middle of the unit cell. The geometry optimization was carried out as a function of the normal stress by minimizing the total energy of the system.

The crystal structure is fully relaxed under a given set of exchange-correlation potential functions and found an equilibrium structure with a minimum total energy. Our calculated value of the lattice constant with GGA-PBE method is obtained as 3.47 Å. and is quite compatible with theoretical value [3] (3.51 Å). The resulting energy versus volume curve is indicated in Figure 2. The equilibrium value of the volume is nearly 280 $Bohr^3$ at the minimum energy value.

* Corresponding author: yasemin@gazi.edu.tr

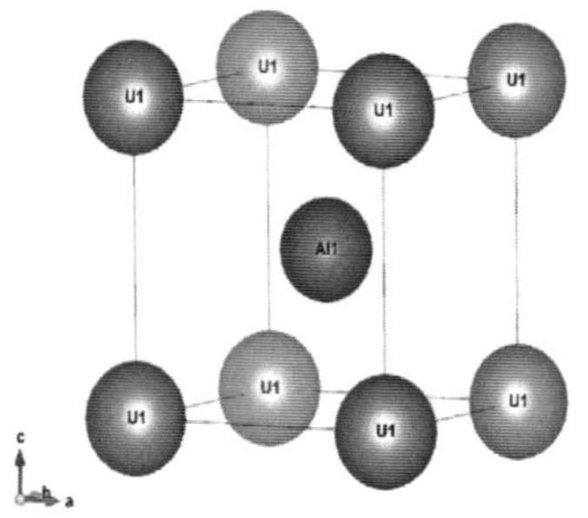

Fig. 1. The crystal structure of UAl alloy in B2

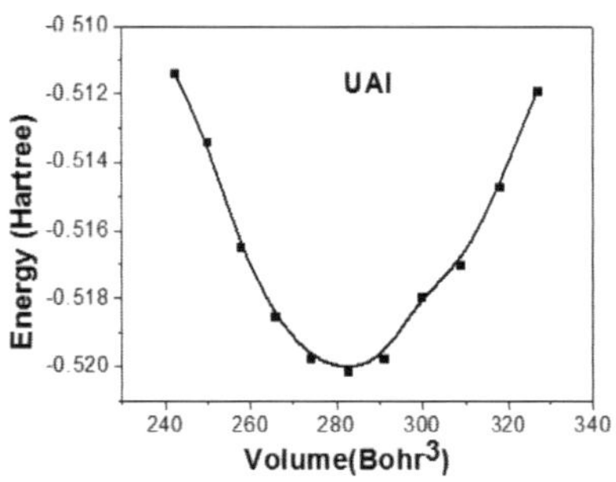

Fig. 2. The energy versus volume graph of UAl crystal structure.

The elastic constants reveal that the response of the crystal to external forces and examine the strength and resistivity of the compounds. For this reason, the second order elastic constants are obtained. To calculate the elastic constants through the first-principles modelling of materials from their known crystal structures, there are basically two common methods [11,12] an approach based on analysis of the total energy of properly strained states of the material in the volume conserving technique and an approach based on the analysis of changes in calculated stress values resulting from changes in the stress-strain technique. In this study, the stress-strain technique is used to determine the second-order elastic constants (C_{ij}) of UAl alloy in CsCl-type crystal structure.

The Born's stability criteria's [13] should be satisfied for the stability of lattice. The known Born's stability conditions of cubic crystals are: $C_{11}>0$, $C_{11}-C_{12}>0$, $C_{44}>0$, $C_{11}+2C_{12}>0$ and $C_{12}<B<C_{11}$ for mechanical stability. According to the Born's stability conditions, UAl alloy is mechanically unstable using the second order constants for calculations.

Table 1. Calculated equilibrium lattice constant (*a*), bulk modulus (*B*), B/G ratio, second order elastic constants, Young's modulus (*E*) and Poisson's ratio of UAl alloy CsCl-type crystal structure* [3]

a (Å)	*B* (GPa)	*B/G*	C_{11} (GPa)	C_{12} (GPa)	C_{44} (GPa)	υ
3.47	88.1	1.14	24.5	120.0	39.0	0.16
3.51*	82*	-1.73*	22.0*	112.0*	35.0*	-2.85*

It is known to be such that the bulk modulus (*B*) determines the resistance of material fracture and the shear modulus (*G*) represents the resistance to plastic deformation. The *B/G* ratio is relevant to ductile/brittle behavior for solids. It is well-known that the critical value is 1.75 and/or greater than that value the material is regarded as ductile [14]. UAl alloy indicates that the brittle behaviour due to the fact that the present value of *B/G* is 1.14 as given in Table 1 at zero pressure.

3.2. Electronic properties

Electronic structure calculations are performed to evaluate the electronic properties of UAl alloy in CsCl-type crystal structure along the high symmetry directions in inclusion of total density of states (DOS) with band structure calculated by GGA scheme as presented in Fig. 3.

It can be seen from Fig. 3 that the UAl alloy exhibits the metallic nature of the compound due to the fact that at around Fermi level (E_F) that is set to the zero band structures are nested. According to the band structure and density of states (DOS), there is no band gap. We also have presented the partial density of states (PDOS) and total density of states (TDOS) of UAl alloy for CsCl-type crystal structure in Figure 4.

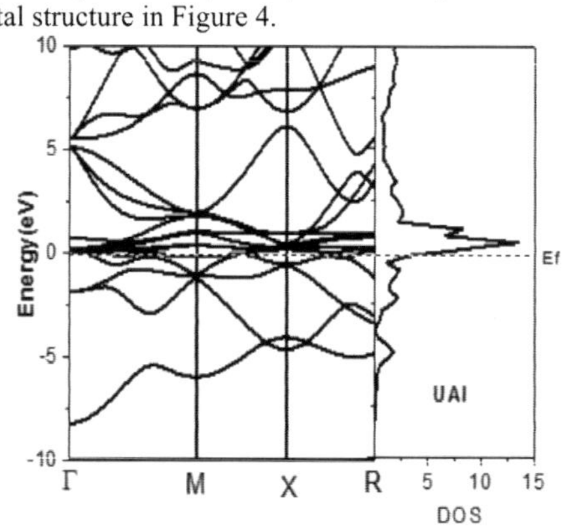

Fig.3. The band structure and density of states of UAl

It is illustrated that in Figure 4, the contribution to the lower valance bands and Fermi energy level are mainly Al-s states and U-d states. The conduction bands are mainly from U-*d* states and partially Al-*p* states. The UAl alloy has a metallic behaviour due to the fact that it has no band gap.

3.3 Vibrational and thermodynamic properties

The vibrational properties determine the lattice dynamical stability using the phonon spectra of the materials. The phonon frequencies of UAl alloy in CsCl structure have been computed using the PHONOPY code [15,16] that is based on the forces obtained from the VASP. The PHONOPY code calculates force constant matrices and phonon frequencies using the density functional perturbation theory as described in Ref. [16]. The present phonon dispersion curves along several high symmetry directions using a 2x2x2 cubic supercell of 16 atoms are illustrated in Figure 5. UAl alloy in CsCl crystal system has two atoms in the unit cell, in the phonon spectrum has six phonon branches, including three acoustic branches and three optical branches. Therefore, it strongly supports the dynamical unstability of this compound in CsCl structure.

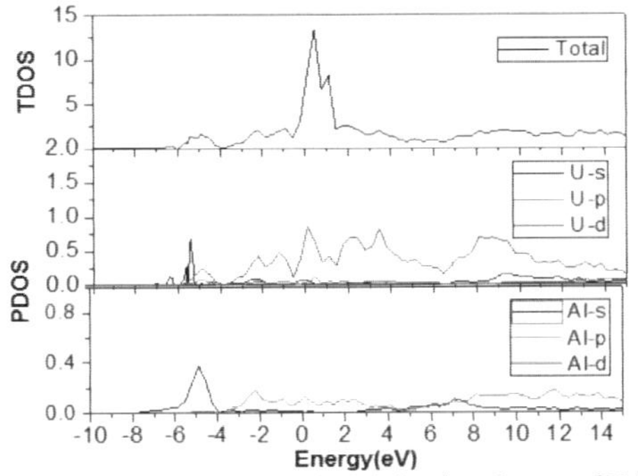

Fig. 4. The partial and total density of states of UAl

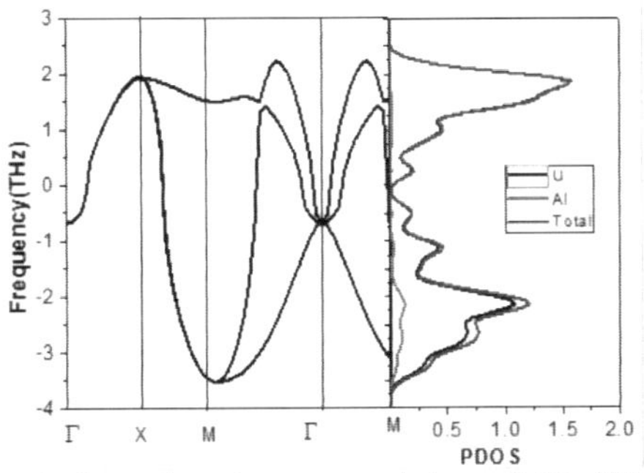

Fig.5. The fonon dispersion curves and phonon DOS of UAl

As shown on the right side of Fig. 5, the phonon density of the states of the UAl alloy consists of two groups of bands. The higher modes with frequencies up to 3.0 THz are composed of the U atoms. In that region, U atoms are more dominant than Al atoms. The modes in the range -3.5 – -0.5 THz are predominantly determined by the vibrations of the U atoms with small contribution Al atoms.

The temperature dependent properties are also estimated through the thermodynamic quantities based on the quasi-harmonic approach and thermal electronic excitation. By using the phonon frequencies, the thermodynamic parameters such as heat capacity (C_v) and entropy (S) have been determined under quasi-harmonic approximation. The contributions from the lattice vibrations to the entropy and the heat capacity of UAl alloy are also given with Fig. 6. To decrease the probable influence of anharmonicity, the temperature is limited to 1000 K. The contribution from the lattice to the heat capacity follows the Debye model and it approaches Dulong–Petit limit at high temperatures.

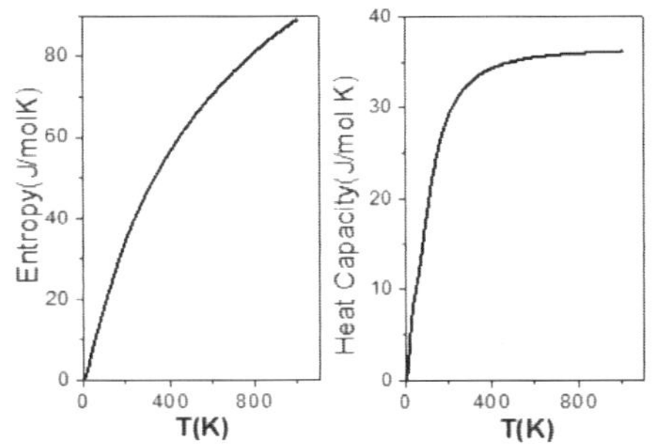

Fig. 6. Phonon thermodynamic properties of UAl

4. Conclusion

In conclusion, we have performed the structural, elastic, electronic, vibrational and thermodynamic properties of UAl alloy using first-principles calculations based on the density functional theory with the frame of GGA-PBE. The calculated lattice constant is in good agreement with the existing result from the other related theoretical study. Second-order elastic constants and other mechanical parameters are obtained and mechanical stability for UAl alloy is investigated using the Born's stability criteria. As a result, UAl alloy is determined as mechanically unstable. UAl alloy exhibits a metallic behaviour at zero pressure. Also, vibrational and thermal properties of UAl alloy is investigated and found that UAl alloy is dynamically unstable. Finally, we have deduced that our theoretical results for basic physical properties of UAl alloy that contributes the applications in materials science for bulk systems.

References

[1] D.D. Keiser Jr, C.R. Clark, M.K. Meyer, Scr. Mater. **51**, 893 (2004)
[2] H.J. Ryu et al., J. Nucl. Mater. **321**, 210 (2003)
[3] S. Kang, T.Gaon, X. Tian, Physica B **407** ,748 (2012)
[4] Ning, Y., Gold Bull., **1**, 38 (2005)
[5] Golyev, B.B., *Synthesis of Alloys*. (Metallurgy Press, Moscow,1984)
[6] Ahmad, S.; Ahmad R.; Jalali-Asabadi S.; Ali Z.; Ahmad I., J. of Magnetism and Magnetic Materials, **422**, 458 (2017)
[7] Kresse, G.; Hafner, J., *Phys. Rev. B*, **49**,14251(1994)
[8] Kresse, G.; J. Furthmuller, J., Comput. Mater. Sci. **6** ,15 (1996)
[9] Kohn, W.; Sham, L., Phys. Rev. A , **40**,1133 (1965).
[10] Perdew, J.P.; Burke, K.; Ernzerhof, M., Phys. Rev. Lett., **77**, 3865 (1996)
[11] Hellman, H., *Introduction to Quantum Chemistry*. (Leipzig Deuticke, 1937)
[12] Monkhorst, H.J.; and Pack, J.D.; Phys. Rev. *B*, **13**, 5188, (1976)
[13] Chao, C.; Luo, H.L.; Duwez, P., J. Appl. Phys., **34**, 7 (1963)
[14] Mehl, J., Phys. Rev. B, **47** ,2493 (1993)
[15] Bannikov, V.V.; Shein I.R.; Ivanovskii, A.L., Phys. Status Solidi (RRL), **1** , 89 (2007)
[16] Johnston, I. et al., *Solid State Physics Simulations, The Consortium for Upper-Level Physics Software*, (John Wiley New York, 1996)

New applications and developments in the neutron shielding

Fatma Aysun Uğur

Osmaniye Korkut Ata University, Department of Physics, Osmaniye, Turkey

Abstract. Shielding neutrons involve three steps that are slowing neutrons, absorption of neutrons, and impregnation of gamma rays. Neutrons slow down with thermal energy by hydrogen, water, paraffin, plastic. Hydrogenated materials are also very effective for the absorption of neutrons. Gamma rays are produced by neutron (radiation) retention on the neutron shield, inelastic scattering, and degradation of activation products. If a source emits gamma rays at various energies, high-energy gamma rays sometimes specify shielding requirements. Multipurpose Materials for Neutron Shields; Concrete, especially with barium mixed in, can slow and absorb the neutrons, and shield the gamma rays. Plastic with boron is also a good multipurpose shielding material. In this study; new applications and developments in the area of neutron shielding will be discussed in terms of different materials.

1 Introduction

Penetrating radiation serves many useful purposes, but it can also damage or destroy living organism, and these effects must be considered when radiation is used. Experience has shown that radiation can be used safely if its dangers are understood and work with radiation is planned to minimize these dangers and eliminate unnecessary exposure. The human sense hearing, sights, taste, smell and touch- cannot detect ionization radiation. Therefore, we must use radiation detection and measuring devices to determine the radiation levels that present in any situation and also, we must apply radiation protection rules.

Ionized radiation has several forms: Alpha, beta, neutron particles and gamma, X-rays. Every kind of energy or mass is caused by excesses (or both) of unstable atoms. To reach a balanced state, they must release this extra energy or mass in the form of radiation [1].

When alpha, beta and neutron are particle type radiations, X and gamma are electromagnetic waves with high energies. Lead or depleted uranium used to be protected from X and gamma. The X-rays are similar to the gamma ray, the primary difference between them is that they originate from the electron cloud [1].

Concrete is the most common material used for shielding in nuclear facilities such as particle accelerators, medical hospitals containing radioactive sources and radiation generating equipment and especially nuclear power stations and storing radioactive waste [2].

The protective properties of concrete against radiation may vary depending on the concrete composites. The additives play an important role primarily in modifying the two properties of concrete; Structural strength and radiation protection capacity [2]. Neutron radiation consists of free neutrons and usually occurs as a result of spontaneous or induced nuclear fission neutrons [1].

They can travel hundreds or even thousands of meters in the air, effectively stopping when they are blocked by a hydrogen rich material such as concrete or water. Neutrons, which usually fail to ionize because an atom is not a direct charge, are often indirectly ionizing because they are absorbed into a stable atom, so they become unstable and increase the likelihood of emitting another ionizing radiation [1].

In fact, neutrons are the only type of radiation that can make other materials radioactive. The type and amount of protective material depends on the radiation dose, the activity of the source and the dose rate. However, there are other factors in choosing protective materials such as manufacturing, cost and weight. It is also very important that the materials used for this purpose are located in the country.

In this context, studies on the absorption of radiation in materials have become an important issue that exists locally and it is therefore desirable to have knowledge of effective materials for neutron and gamma ray shielding. Many researchers have reported that concrete is an indispensable material for particle accelerators, medical hospitals, radioactive sources and radiation producing equipment, and in particular for nuclear power plants and radioactive waste repositories. Radiation shielding properties of concrete may vary depending on the concrete composites [2].

Composites play an important role in essentially changing the two properties of concrete structural strength and radiation protection capacity.

* Corresponding author: aysunugur@osmaniye.edu.tr

Thus, linear attenuation coefficients increase as the Z of the absorber increases because photoelectric interactions are increased in high Z materials especially for low energy photons [2].

Because of the high Z effect, lead and concrete are often used to draw the walls of the X ray chambers and sites, and minerals such as hematite and barite are incorporated into the concrete to enhance its effectiveness as a photon shield.

Another aspect of the concrete popularity is the hydrogen content for the neutron shield. Hydrogen in concrete is known to be a radiation protection (especially neutron) process. Concrete blocks absorb neutrons due to hydrogen content. There are several studies about radiation shield processes using concrete; [3-7]. Also, neutron dose transmission values determined for several new concrete types including colemanite and neutron shielding properties of some boron bearing compounds determined and calculated [8,9].

2 Results and Discussions

2.1 Neutron shielding takes place mainly in three stages

1) Neutron slowdown; Neutrons classified according to their energy; E > 100 MeV: High energy neutrons, 10-20 MeV > E > 100-200 keV: Fast neutrons, 100 keV > E > 0.1 eV: Epithermal neutrons, E ~ kT ~ 1/40 eV: Thermal / Slow Neutrons, E ~ meV: Cold and ultra-cold neutrons. Neutrons are slowed down by thermal energy with hydrogenous materials. These materials are usually water, paraffin and plastic. There are some difficulties in using these materials. Water can evaporate or flow, paraffin can burn, plastic is quite expensive. Very fast neutrons can be slowed down by putting lead or iron barriers in front of them.

2) Absorption of neutrons; As mentioned above, the absorbing of neutrons in hydrogenated materials is highly effective.

The neutron capture effect section of H-1 is 0.33 barn. The cross section of the neutron absorption effect of boron is a very high material. The only negative situation when using boron is to emit a gamma ray at low energies. Lead or iron material is used to slow down very fast neutrons. Lead is a very important material for shielding with large atomic number, dense and soft structure.

3) Absorption of gammas due to neutron interaction; Gamma rays are produced by neutron (radiation) retention on the neutron shield, inelastic scattering, and decay of activation products.

Neutron shielding calculations are generally done by computers. There is a classical calculation formula for mono-energetic photons. This equation was defined by Schaeffer in 1973. The protective material is placed in front of the neutrons for slow neutron neutrons, like steel. At the back of this material must be a material containing hydrogen at least 50 cm thick (Fig. 1).

$$\dot{D} = \dot{D}_0 e^{-\Sigma_R t} \quad (2.1)$$

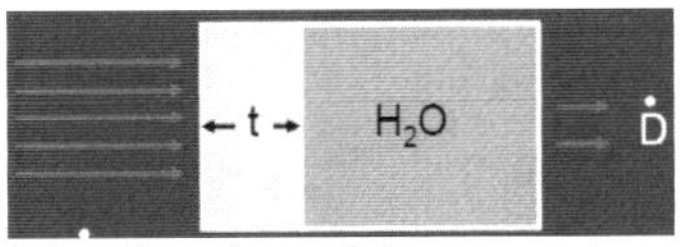

Fig. 1 Experimental Arrangement

$\dot{D}$ is the dose rate with shield, $\dot{D}_0$ is the dose rate without shield, t is the shield thickness(cm)

Σ_R is the neutron removal cross section(cm-1). If these calculations are made when an Am-Be neutron source is used, the building factor is approximately 5. When Equation 2.1 is regulated, the structure factor has the following effect in the equation.

$$\dot{D} = \dot{D}_0 e^{-\Sigma_R t} \quad (2.2)$$

B is the building factor in Eq. (2.2). The material removal cross sections from NBS Handbook 63 are given in Table 1.

Table 1. Removal Cross Section of Material

Material	Removal Cross Section Σ_R (cm^{-1})
Water	0.1030
Iron	0.1576
Ordinary concrete	0.0942
Barytes concrete	0.0945
Graphite	0.0785

For what it is worth, the removal cross section is approximately 2/3 to 3/4 of the total cross section. In neutron shielding calculations, we might also use the mass attenuation coefficient symbolized ΣR/ρ. According to Schaeffer (1973), the mass attenuation coefficient (ΣR/ρ) for fast neutrons can be approximated with your choice of one of the following: [1].

$$\Sigma R / \rho = 0.19Z - 0.743cm^2 / g(Z \leq 8) \quad (2.3)$$

$$= 0.125Z - 0.565cm^2 / g(Z > 8) \quad (2.4)$$

$$\Sigma R / \rho = 0.206A - 1/3Z - 0.294 \approx 0.206(AZ) - 1/3 \quad (2.5)$$

$$\Sigma R / \rho = 0.21A - 0.58 \quad (2.6)$$

The previous equation can be written

$$\dot{D} = B \dot{D}0 \, e^{-(\Sigma R / \rho)\rho t} \quad (2.5)$$

The lower energy of the gamma rays or neutrons is easier if it is to protect them. The more various high-energy gamma energies a radioactive source emits, the greater the need for shielding. For gamma rays, the higher the atomic number of the shielding material, the greater the weakening of the radiation.

For neutrons, the material allows the radiation to attenuate as the atomic number of the shielding drops. The greater intensity of both radiation and shielding material means the greater weakening of radiation. Although water and air have essentially the same effective atomic number, condensation of water causes a better shield.

Lead is a good shielding material with a large atomic number (Z = 82), high density (11.34 gr/cm^3), easily shaped, melted and relatively inexpensive. However, it is disadvantageous that it is weak, toxic, that the new lead material is obtained from natural lead and the recycling is contaminated. Lead can also be produced in a variety of forms. For example; standard bricks, interlocking bricks, sheets, tape, wire, pipe. It can be used in tungsten alloy as shielding material. Tungsten alloy contains iron, copper, cobalt and nickel. Tungsten has a high atomic number (74). It's very density (14-18 g/cm^3). It is strong, it is not toxic. It can be made finer than the lead. However, shaping and manufacturing are difficult and expensive at the same time. Another material used in shielding is uranium, which is enriched with high atomic number and density, as well as mechanically strong formation. The disadvantages of this material are expensiveness, radioactivity and its ability of quickly inflame. One of the best materials used in neutron shielding is water because it is even more homogeneous than the best shielding material. At the same time, it is transparent, inexpensive, radioactive sources and equipment can be installed in the water. Source leaks can be detected by sampling water. As the water shielding material, the disadvantages easily flow and evaporate. In addition, a leaking source can easily contaminate water.

Concrete is an important material to shielding both neutrons and gammas. Barium can be added to make it more effective. Very thick walls may be needed. It is difficult to change them.

3 Conclusions

For radiation sources emitting neutron and gamma, materials such as water, concrete, mixed concrete with barium, lead tunnels can be used as shielding material. These materials have certain advantages or disadvantages and they can be used according to the energy of radioactive materials and the type of radiation they emit.

References

1. www.mirion.com
2. O. Gencel, Annals of Nuclear Energy 38, 12, December 2011, Pages 2719-2723.
3. K. Shin, H. Hirayama, Radiat. Phys. Chem., 61 (2001), pp. 583-584
4. A.M. Hathout, Radiat. Phys. Chem., 44 (1994), pp. 115-117
5. W.A. Kansouh, Radiat. Phys. Chem., 44 (1994), pp. 164-167.
6. Y. Harima, Radiat. Phys. Chem., 41 (1993), pp. 631-672.
7. A.M.I. Kany, M.I. Elgohary, S.M. Kamal, Radiat. Phys. Chem., 44 (1994), pp. 157-160.
8. T. Korkut et al., Ann. Nucl. Energy, 37 (2010), pp. 996-998.
9. T. Korkut, A. Karabulut, G. Budak, H. Korkut, Radioanal. Nucl. Chem., 286 (2010), pp. 61-65

Depth dose calculations

Fatma. Aysun Uğur[1,*], *Elif* Gören[2]

[1]Osmaniye Korkut Ata University, Department of Physics, Osmaniye, Turkey
[2] Cukurova University, Department of Physics, Adana, Turkey

Abstract. In this study, calorimetric method is used to derive the depth doses calculation. In the experiment at the KeV level, an Ir-192 source with three different gamma energies was used. The source activity is approximately 9.15 curie. The Ir-192 radioactive source is mainly used for industrial purposes. This study had been made to determine organ dose values in any industrial accident. The data are calculated as surface dose and depth dose. Values for different time intervals have been measured. Time and depth parameters for measurements and calculations were obtained for 30 different values. The results of the experiment and the calculations have been compared.

1 Introduction

Industrial radiography is a frequently used method of non-destructive testing for many industrial products. Industrial radiography is a routine operation requiring high radiation levels and relying primarily on human diligence in observing safety procedures to prevent accidents [1].

Ir-192 source is mostly used in industrial radiography. Ir-192 is normally produced by neutron activation of natural-abundance iridium metal, usually in nuclear reactors. The strength (or specific activity) of a resulting Ir-192 is related to the amount of neutron irradiation and length of time to which the natural-abundance iridium metal is exposed. During irradiation only the stable isotope Ir-191 is activated to produce Ir-192 by absorbing a neutron. Radioactive Ir-192 has a half-life of 73.83 days. Radioactive Ir-192 is used principally for non-destructive testing (NDT) and, to a lesser extent, as a radio-tracer in the oil industry. Industrial gamma radiography involves the testing and grading of welds on pressurized piping, pressure vessels, high-capacity storage containers, pipelines, and certain structural welds. Other tested materials include concrete (locating rebar or conduit within the concrete), machined parts, plate metal, and pipe wall. Gamma radiography is also used to identify flaws in metal castings and welded joints, as well as to indicate structural anomalies due to corrosion or mechanical damage [2].

Literature studies have shown that industrial radiography also has various accidents. Workers working with radioactive material in these incidents are exposed to radiation. Experiments are carried out using the calorimetric method to detect these radiation doses. In addition, organ doses taken by people who have been caught in the accident can be calculated by this method. For this purpose, depth dose calculations were made using water phantom and a farmer dosimeter. In this study, Ir-192 was used as a radioactive source. In this study, experiments were carried out in TAEK Cekmece Nuclear Research Centre Health Physics Department SSDL (Secondary Standard Dosimetry Laboratory).

2 Material and methods

In the field of ionizing radiations, the absorbed dose is one of the basic quantities commonly used in dosimetry. The main reason of that it is closely related to the biological effects of radiation. Water is of special interest as an absorbing material because it is very similar to human tissue. One of the duties of the national standards laboratories is to establish accurate standards for absorbed dose in water. To this end, much effort has been put into the development of various experimental approaches. One of the most frequently used ones is the calorimetric method because its basic principle is simple [3].

Therefore, the measurements made here were obtained in detail by this method. In these measurements, the values of pressure and temperature were recorded regularly and added to the calculations as correction parameters [1].

2.1. Depth dose calculations

The aim of the present work was to measure the depth dose rates by using a water phantom and 192Ir source to obtain experimental and theoretical results. The absorbed dose (D_d) at depths from 0 to 35 cm in tissue with the source in surface contact was calculated by;

$$D_d = \frac{0{,}96 A_2}{\left(d + 0{,}24\right)^2} f\left(d\right) \qquad (1)$$

* Corresponding author: aysunugur@osmaniye.edu.tr

where, An equal the source activity in unit of curie (A is the exposure rate at 1 cm), Γ is the exposure rate constant, 0.96 is the conversion factor to rads/Roentgen, d is the tissue depth in unit of cm, 0.24 is the estimated encapsulation thickness in unit of cm, f(d) is the scatter and attenuation polynomial of Shalek and Stovall [4] for d ≤ 15 cm or exp(-0.035d) for d ≥ 15 cm.

Since there is a likelihood that the source may have been at a distance of up to 2 cm from the surface, depth doses were also calculated for these circumstances using standard depth-dose improvement factors with greater distances from the surface. The results are shown in Fig. 3.

2.2 Experimental Arrangement

A schematic view of the experimental arrangement is given in Fig. 1. The Ir-192 source is 2 cm diameter and 0.56 cm in length. On the day of the experiment the source activity is 9.15 Ci. The half-life of the source is 74.2 days. Ir-192 is a meta-stable isotope of Iridium emitting gamma radiation with a low level of Linear Energy Transfer (LET) and high penetration in human body. This isotope frequently has been used for industrial radiography [5, 6].

The presence of impurities, especially those with long half-lives and high radiation energies, can significantly alter the absorbed doses. Its energy spectrum is extremely complex, because it has got 30 gammas and 40 X-rays [7].

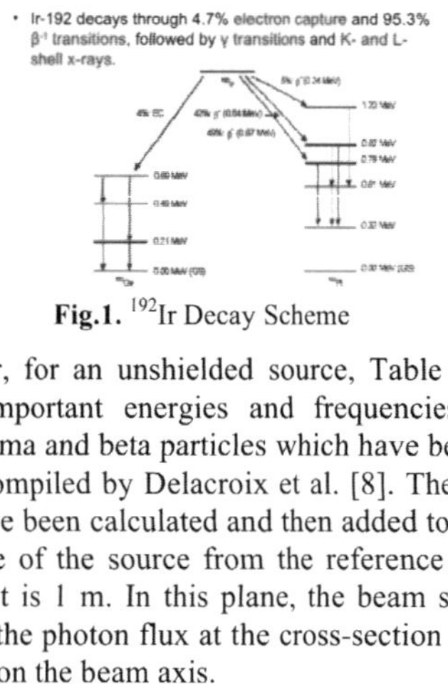

Fig.1. ^{192}Ir Decay Scheme

However, for an unshielded source, Table 1 shows the most important energies and frequencies of the emitted gamma and beta particles which have been taken from data compiled by Delacroix et al. [8]. The average energies have been calculated and then added to Table 1. The distance of the source from the reference plane of measurement is 1 m. In this plane, the beam size is 10 cm×10 cm; the photon flux at the cross-section border is 50% of that on the beam axis.

Table 1. Gamma and beta energy distribution for an unshielded ^{192}I source [9]

^{192}I	Gamma E(keV)	Frequency (%)	Beta E(keV)	Freuency (%)
E1	317	83	256	6
E2	468	48	536	41
E3	604	8	672	48
Average	372		583	

The water phantom consists of a cubic perspex tank, of side 30 cm, with walls 14 mm thick. The wall facing the beam has a thickness of 4.0 mm (0. 476 g.cm^{-2}) over a section of 15 cm×15 cm to reduce the correction due to the non-equivalence of the perspex front face with water. The phantom, filled with demineralized water, is positioned in such a way that the reference plane is at a depth equivalent to 5g.cm-2 in water [9].

The ionization chamber, inserted into a perspex support (1.85 mm thick), can be moved inside the water along the beam axis by means of a translation table fixed on the phantom walls. This allows an accurate adjustment of the chamber position

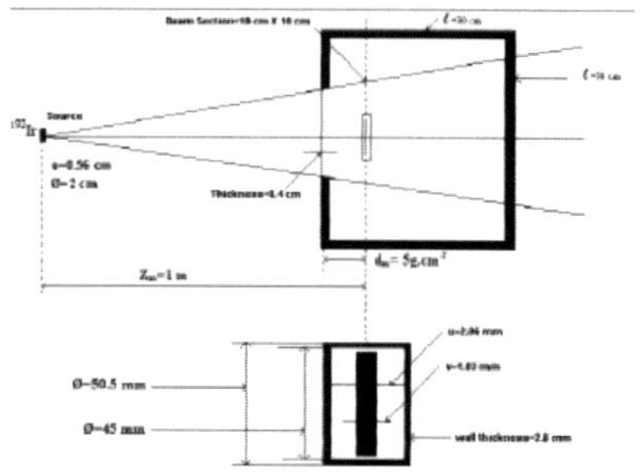

Fig.2. Experimental Arrengement

To avoid long-term variation of the ionization current due to the possible influence of humidity on the ionization chamber, the phantom is filled up and drained every day. The temperature of the chamber is measured with a calibrated thermistor. To match the air conditions of the chamber cavity, the thermistor is placed in air inside a Perspex tube dipped into water. The temperature difference between the positions of the thermistor and of the chamber has been checked and found to be small (less than 0.01°C), so the error introduced is negligible [1].

A Baldwin- Farmer 0.6 cc ionization chamber with a ±%3 calibration accuracy (traceable to the National Bureau of standards) was exposed, in air, at a distance of 14 cm from the source. The positioning error was estimated to be less than ± 0.5 cm. This measurement indicated an average exposure rate on 15 June 2016, of 224.06 R/h at distances of 14 cm. The standard deviation of eight such measurements was less than %3. This corresponds to a source activity of 9.15 ± 0.65 Ci (0.33 ± 0.02 TBq), with the uncertainty reflecting possible positioning errors. This activity determination was based on an exposure rate constant of 4.8 R-cm^2/mCi-h as reported in NCRP (National Council on Radiation Protection and Measurements) Pamplet 40 [7,10]. Although estimates of the exposure-rate constant for ^{192}Ir range from 4.3 to 5.0 R-cm^2/MCi-h, the value of 4.8 was selected to be consistent with the exposure and dose calculations of this paper which are based on the actual exposure measurements and are independent of the exact source activity [7,1].

2.3 Experimental and its calculation

Firstly, perspex tank has been filled with demineralised water. ^{192}Ir source has been placed at a distance of 3 cm from the tank. Ionization chamber has been placed at a distance of 2.5 cm. Then, radiation count per minute has been carried out at this position. The exposure unit is generally expressed in unit of Ront/h. However, this expression can be transformed into rad/h by multiplying with a factor of 0,96 rad/Ront as given below:

$$Exposure unit = \frac{Rönt}{h} x 0{,}96 \frac{rad}{Rönt} = \frac{rad}{h} \qquad (2)$$

Using the proper conversion factors an expression for absorbed dose in the unit of rad/h is found as

$$D = Count\left(\frac{Rönt}{h}\right) x CP x TP x 0{,}96 x 60 \qquad (3)$$

In this equation; calibration factor of ionization chamber (CP) and temperature-pressure factor (TP as also calculated from equation [11] below) were 1.004 and 1.0165, respectively

$$TP = \frac{273{,}16\,^{0}K + 23{,}5\,^{0}C}{293{,}6} x \frac{1013}{1008{,}4} = 1{,}01655 \qquad (4)$$

Here, the laboratory temperature on the day of the experiment was taken as 23.5°C and pressure as 1008.4 mb. Furthermore, normal temperature and air pressure on sea level were 20°C and 1013 mb, respectively.

Then the water phantom was irradiated for 20 minutes with a change between SSDL = 100 and 200 cm and d = 20-50 cm. The results are shown in Fig. 3.

3 Results and discussion

The purpose of this study was to determine the depth dose. The distance between the radioactive source water phantom is 100 to 200 cm. The depth has been changed from 20 to 50 cm. When the results aren't obtained graphically, the depth-dependent dose change values change exponentially. However, at some points deviations have been observed due to back scattering. When the results are carefully analysed, the changes obtained when the gamma rays emitted from the radioactive source interact with the substance overlap with the classical approach.

Gamma rays interfere with the matter, resulting in photoelectric effect, Compton scattering and pair production (Fig. 4). Fig. 3 is similar to Fig. 5 when viewed.

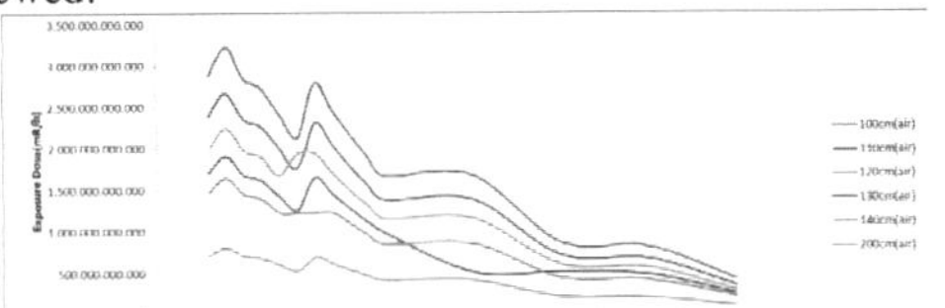

Fig.3. Irradiation dose change for Ir-192 radioactive source

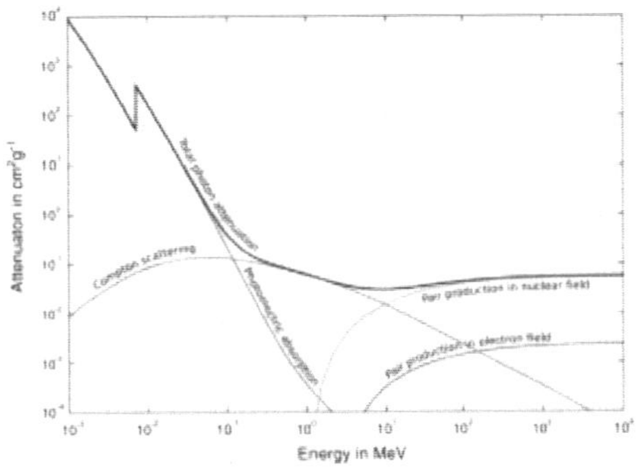

Fig.4. The relative importance of various processes of gamma radiation interactions with matter

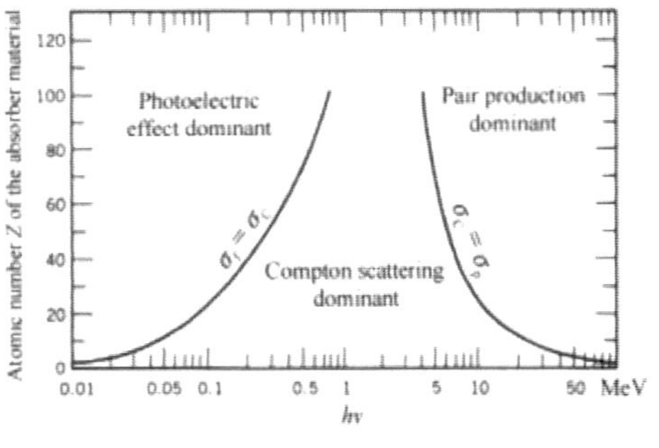

Fig. 5. The relative importance of various processes of gamma radiation interaction with matter.

At low energies, the photoelectric effect and the Compton scattering dominate, while the high energies suppress pair production. (Fig. 5) The linear absorption coefficient for water varies between 0,167 and 0, 0706 for gamma rays between 100 KeV and 1000 KeV.

4 Conclusions

Gamma rays interaction with matter is important from the perspective of shielding against their effect on biological matter. They are considered as ionizing radiation whose scattering by electrons and nuclei leads to the creation of a radiation field containing negative electrons and positive ions. The main modes of interaction of gamma rays with matter are the photo effect both in its photoelectric and photonuclear forms, Compton scattering and electron positron pair production. To a minor extent, photo-fission, Rayleigh scattering and Thomson scattering also occur. Gamma rays have both material attenuation (effect of shielding) and geometric attenuation when interacting with matter. There is also the effect of scattering, which is found in a shielding material and called build up effect. The graph obtained for these reasons is similar to the graph of the interaction of gamma rays with matter.

References

1. Uğur et al., Fizika , Volume XVI, Number 2, 516-519 (2010)
2. http://www.isoflex.com/iridium-192
3. M. Boutillon and A.M. Perroche, Phys. Med. Biol., 38 439-454, (1993).
4. R.J. Shalek, M. Stoyall, New York, Academic Press, (1969)
5. A. Jalil, M.A. Molla Health Phys. 62(1): 74-76, (1992)
6. S. Milacic, J. Simic, J. Kobe Med. Sci., 54, 2, E108-E113, (2008)
7. H. F. Eugene and L. B. William, The Medical Basis for Radiation Accident Preparedness, 265-277, (1980)
8. D. Delacroix, J. P. Guerre, P Leblanc, C. Hiykman, Radionuclide and Radiation Protection Data Handbook 2002 Radiat. Prot. Dosim. 98 1-168, (2002)
9. M. Boutillon and A.M. Perroche, Phys. Med. Biol. 38 439-454, (1993)
10. Protection Aganist Radiation From Brachytherapy Sources. NCRP Pamphlet No.40, National Council on Radiation Protection and Measurements, Marh I, (1972)
11. Turkish Atomic Energy Authority website (www.taek.gov.tr)

Material dependence of ^{2}H(d,p)^{3}H cross section at the very low energies

Ali İhsan Kılıç[1], *Konrad* Czerski[2,3], *Fadime* Kuştan-Kılıç[2,4], *Natalia* Targosz-Sleczka[2], *Daniel* Weissbach[2,3], *Armin* Huke[3], *and Götz* Ruprecht[2]

[1]University of Guelph, 50 Stone Road E, Guelph, Canada
[2]University of Szczecin Department of Physics, Szczecin, Poland
[3]Institute for Nuclear Solid-State Physics, Berlin, Germany
[4]University of Gazi, Department of Physics. 06500 Teknikokullar, Ankara, Turkey

Abstract. Calculations of the material dependence of ^{2}H(d,p)^{3}H cross section and neutron-to-proton branching ratio of d+d reactions have been performed including a concept of the 0^+ threshold single particle resonance. The resonance has been assumed to explain the enhanced electron screening effect observed in the d+d reaction for different metallic targets. Here, we have included interference effects between the flat and resonance part of the cross section, which allowed us to enlighten observed suppression of the neutron channel in some metals such as Sr and Li. Since the position of the resonance depends on the screening energy that strongly depends on the local electron density. The resonance width, observed for the d+d reactions in the very hygroscopic metals (Sr and Li) and therefore probably contaminated by oxides, should be much larger than for other metals. Thus, the interference term of the cross section depending on the total resonance width provides the material dependences.

1. Introduction

The ^{2}H(d,p)^{3}H and ^{2}H(d,n)^{3}He reactions have been investigated from beginnings of physics [1]. The d+d reactions are especially important for developing the next generation of power plants [2], understanding of the nuclear process within stars [3], exploring the physics of few-body nucleon systems [4], solving some astrophysics problems [5] and the problems of nucleon nucleon–interaction [6].

The exponential-like increase of the reaction cross-section observed for decreasing projectile energies, as compared to the cross section for bare nuclei, can be described corresponding to the reduction of the Coulomb barrier height by a screening energy. The theoretical values of the large screening energies are by at least a factor of two smaller than experimentally determined [7]. The enhanced electron screening effect recently observed in the d+d fusion and some other nuclear reactions in the metallic environments is essential for general understanding of stellar reaction rates white and brown dwarfs or giant planets since it can strongly increase the nuclear reaction rates in dense astrophysical plasmas [8]. The angular distributions and the branching ratio between the ^{2}H(d,p)^{3}H and ^{2}H(d,n)^{3}He reactions strongly depend on the target material, being different self-implanted deuterized metallic environments. Whereas for metallic targets made of Al, Zr, Ta both observable show the same projectile energy dependence as in the case of the gas target measurements, for Sr and Li a significant suppression of the neutron channel and increase of its angular anisotropy could be observed for deuteron energies below 20 keV [9].

To describe these effects, we applied an alternative model based on the T-matrix elements [10] which corresponds to broad ^{4}He resonance contributions. Within this approach, interference effects between resonances could be easily included. The theoretical results agree very well with experimental data obtained for usual materials and width multichannel R-matrix theory [11]. To explain the experimental data obtained for unusual materials, it was necessary to include a new 0^+ threshold resonance. This resonance should have a single particle deuteron-deuteron structure and be very narrow (the total width of few eV). The contribution of the 0^+ resonance to the reaction cross section could reduce large screening energies observed.

2. Single particle 0^+ threshold resonance

A hypothetical 0^+ resonance in ^4He placed very close to the d+d reaction threshold has been proposed [12]. The high energy tail of the resonance would contribute to the observed enhancement of the cross-section due to the screening effect and reduce experimental screening energy down to the theoretical value the mathematical point of view a coherent sum of two reaction amplitudes reads as follows:

$$\left|A_1 e^{i\emptyset_1} + A_2 e^{i\emptyset_2}\right|^2 = A_1^2 + A_2^2 + 2A_1A_2\cos(\emptyset_1 - \emptyset_2) \quad (1)$$

Where $\emptyset_1$ and $\emptyset_2$ are nuclear phase shifts. The above procedure can be applied to the coherent sum of cross sections resulting from the narrow threshold 0^+ resonance (σ_R) and from broad overlapping resonances giving a flat cross section (σ_F) contribution.

Interferences in the total cross section are possible only between resonances of the same spin and parity. Thus, the interference part of the total cross section relates only to 0^+ resonances, both a narrow postulated one and the broad resonances of the flat contribution. Therefore, since cross section in this consist of two components, it can be written as

$$\sigma(0^+) = \sigma_F(0^+) + \sigma_R + 2\sqrt{\sigma_F(0^+)}\sqrt{\sigma_R}\cos(\emptyset_1 - \emptyset_2) \quad (2)$$

$\sigma_F(0^+)$ corresponds to α_0 transition matrix elements [10]. Since the Deuteron energy of about 10 keV is much larger than the resonance energy and the resonance width, expression for the resonance phase shift can be simplified

$$\tan\emptyset_R = \frac{\Gamma}{2(E-E_R)},\quad E \gg E_R \quad E \gg \Gamma \quad \emptyset_R = \frac{\Gamma}{2E} \ll 1 \quad (3)$$

therefore

$$\cos(\emptyset_F - \emptyset_R) = \sin\emptyset_F \sin\emptyset_R + \cos_F\cos_R \approx \frac{\Gamma}{2E}\sin\emptyset_F + \sqrt{1 - \frac{\Gamma^2}{4E^2}}\cos\emptyset_F \quad (4)$$

expanding the square root value to the first term in Taylor series we get:

$$\cos(\emptyset_F - \emptyset_R) \approx \frac{\Gamma}{2E}\sin\emptyset_F + \left(1 - \frac{1}{2}\frac{\Gamma^2}{4E^2}\right)\cos\emptyset_F \approx \frac{\Gamma}{2E}\sin\emptyset_F + \cos\emptyset_F \quad (5)$$

the final expression was obtained by neglecting the quadratic term. Since the $\frac{\Gamma}{2E} = 10^{-4}$ for E=10 keV we can simplify the expression further

$$\cos(\emptyset_F - \emptyset_R) \approx +\cos\emptyset_F \quad (6)$$

total cross section is given by

$$\sigma_{Tot} = \frac{1}{\sqrt{E}}\frac{1}{\sqrt{E_G}}P(E)S(E) \quad (7)$$

where P and S denote the s-wave penetration factor and the astrophysical S-factor, respectively. The total S-factor can be expressed by a sum of S_a and S_b arising from the isotropic and an isotropic contribution. Roughly, $\frac{1}{3}$ of the L=0 partial cross section (corresponding to S_a) comes from the 0^+matrix element σ_0. Therefore, the 0^+ flat cross section contribution is given by

$$\sigma_F(0^+) = \frac{1}{3}\sqrt{\frac{1}{E_{cm}E_G}}\,\mathrm{P(E)S_a} \quad (8)$$

here, E_G represents Gamow energy, equal for 986 keV for the d+d system. E is the energy of the deuteron in the center of mass system, and can be expressed by the wave number k, we get:

$$\sigma_F(0^+) = \frac{1}{3}\frac{\pi}{k}\mathrm{P(E)}\frac{\sqrt{2m}}{\hbar}\frac{1}{E_G}S_a \quad (9)$$

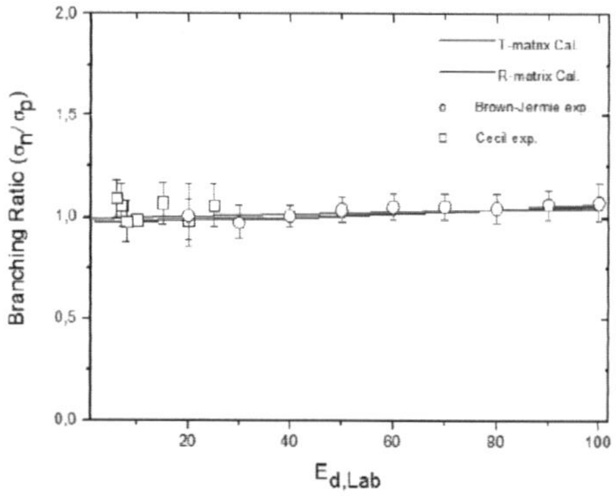

Fig. 1. Experimental neutron-to-proton branching ratio obtained for targets of Ta, Sr, and Li [9] compared to the calculations assuming deuteron polarization (blue-line), and without polarization(red line)

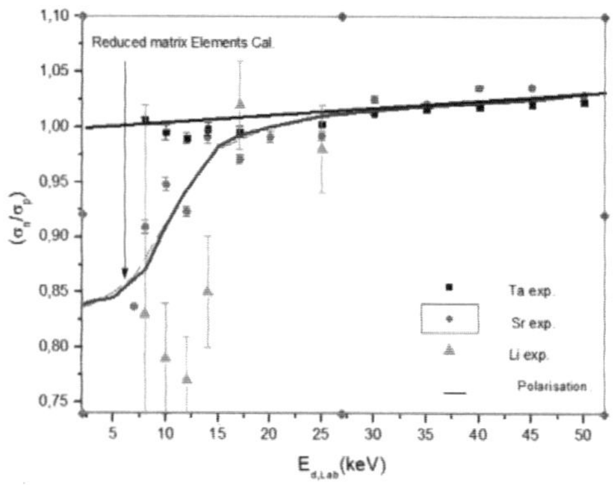

Fig. 2. Experimental and theoretical neutron-to-proton branching ratio. The experimental data are from [16] and [17].

For the resonance contribution, we apply Breight-Wigner formula of the proton channel (similar for the neutron channel)

$$\sigma_R = \frac{\pi}{k^2}\frac{\Gamma_d\Gamma_p}{(E-E_0)^2+\frac{\Gamma^2}{4}} \approx \frac{\pi}{k^2}\frac{\Gamma_d\Gamma_p}{E^2} \quad (10)$$

The d+d reaction cross section can be enhanced by a hypothetical 0^+resonance state in He^4 compound nucleus. Due to its molecular structure [13], it can be described as a single-particle resonance width:

$$\Gamma_d(E) = 2k_d a P(E_d)\frac{\hbar^2}{\mu a^2}|\Theta| \quad (11)$$

The d+d reaction cross section can be enhanced by a hypothetical 0^+resonance state in He^4 compound nucleus. Due to its molecular structure [13], it can be described as a single-particle resonance width:

Here, a=7fm is the reaction channel radius. μ is reduced mass of deuteron, $|\Theta|^2$ stays for the dimensionless reduced width being equal to unity for the single- particle resonance. Taking account that E $\gg E_R$, $\Gamma \gg \Gamma_d$ and we put the Γ_d to σ_R.

Finally, the resonance cross section reads as follows:

$$\sigma(0^+) = \frac{\pi}{k}P(E)[A^2\frac{1}{3}S_a + \frac{B\Gamma_p}{E^2} + 2A\sqrt{\frac{S_a}{3}}\sqrt{\frac{B\Gamma_p}{E}}\cos\emptyset_F \quad (12)$$

Where

$$A^2 = \frac{2}{3\pi}\frac{\sqrt{m}}{\hbar}\frac{1}{E_G}, \quad B=2\frac{2\hbar^2}{\mu a} \quad (13)$$

Due to the destructive interference effects, we can now reduce σ_0 matrix elements [10] in the low energy region as it is necessary to explain the observed material dependence of the branching ratio Fig. 1. The material dependence can be additionally enhanced assuming a smaller neutron partial width of the resonance. The branching ratio can be calculated using formula 12. for neutron channel and proton channel separately. In usual case, Γ_n, Γ_p are much smaller than for Sr and Ta. Therefore, branching ratio can be described without resonance contribution (Fig. 2).

3. Conclusions

In the present work, we could show, how the material dependence of the branching ratio of the d+d fusion reaction might be explained applying the concept of the threshold resonance. The branching ratio does not depend on the penetration factor any more. The screening effect should not directly influence its value. In the usual case, Γ_n and Γ_p also do not depend on the target material as they are only proportional to the nuclear wave functions. The destructive interference could be also observed in the energy dependence of the enhancement factor obtained in the new UHV experiment on the Zr target [14]. The resonance contribution reduces the estimated screening energy by a constant factor, making it closer to the theoretical value. The direct resonance evidence can be expected by the precision measurements of the d+d reactions at deuteron energies below 5keV where different energy dependence of the enhancement factor due to electron screening and the resonance could much more visible. Corresponding experiments are planned at the University of Szczecin applying the new accelerator facility with a UHV target chamber [15].

References

1. E. Lawrence et al., Phys. Rev. Lett.44, 56 (1933).
2. R.M. More, Phys. Rev. Lett 51 **396** (1983)
3. E. Margaret et al, Rev. Mod. Phys. 29 **547** (1957)
4. S.E. Koonin and M. Mukerjee, Phys. Rev. **C42** (1990) 1639.
5. D. Kirkman et al., Apj S, 149 **1.6** (2003) 1639.
6. C. Angula et al., Z. Phys. **A345** (1993) 231.
7. A. Huke et al. Phys. Rev. C **78** (2008) 015803
8. Zs. Faulaop et al., Nucl. Phys. A **758** (2005) 697
9. A. Huke et al. Eur. Phys. J **A 27** (2006) 187-192
10. H. Schieck. Eur. Phys. J **A 44** (2010) 321354
11. H.S. Bosch, G.M. Hale. Nucl. Fusion **32** (1992) 1919
12. N.Targozh-Sleczka et al. Acta Phys. Pol. **V42** (2011) 795-798
13. W.Von Oertzen Z. Phys. A 37 **A44** (1996) 354
14. K. Czerski et al. J. Phys.G, Nucl. Part. Phys **35** (2008) 014012
15. M. Kaczmarski et al. Acta Phys.Pol. B 45 (2014) 509.
16. R.E. Brown and N. Jarmie Phys. Rev. C 41. (1990) 1391.
17. F.E. Cecil et al. Phys.Rev C **47** (1987) 1178

DOI: 10.1051/epjconf/201715401025

The calculation of neutron flux using Monte Carlo method

Mehtap Günay*, *Hilal* Bardakçı

İnönü University, Science and Art Faculty, Physics Department, Malatya, Turkey

Abstract. In this study, a hybrid reactor system was designed by using 99–95% $Li_{20}Sn_{80}$ + 1-5% RG-Pu, 99–95% $Li_{20}Sn_{80}$ + 1-5% RG-PuF_4, and 99–95% $Li_{20}Sn_{80}$ + 1-5% RG-PuO_2 fluids, ENDF/B-VII.0 evaluated nuclear data library and 9Cr2WVTa structural material. The fluids were used in the liquid first wall, liquid second wall (blanket) and shield zones of a fusion–fission hybrid reactor system. The neutron flux was calculated according to the mixture components, radial, energy spectrum in the designed hybrid reactor system for the selected fluids, library and structural material. Three-dimensional nucleonic calculations were performed using the most recent version MCNPX-2.7.0 the Monte Carlo code.

1 Introduction

The waste management required as a result of fuel burning to generate energy from traditional nuclear reactors is one of the most important problems today. For this reason, hybrid reactor system has been developed where fusion and fission occur at the same time. Hybrid reactor is a system developed to obtain more energy and nuclear fuel, decrease the waste amounts in storage areas, recover transuranic elements in used fuels through reprocessing, and render fission products harmless [1-5].

Reactor-grade (RG) plutonium is found in spent nuclear fuel. Isotopic distribution of RG-Pu has composition 1% ^{238}Pu, 62% ^{239}Pu, 24% ^{240}Pu, 8% ^{241}Pu and 5% ^{242}Pu [6]. In this study, $Li_{20}Sn_{80}$ molten salt was used, which has low melting temperatures and low vapor pressure, to acquire sufficient tritium breeding. Beryllium (Be) is used as a neutron multiplier [7]. In this study, a Be zone with a thickness of 3 cm was used between the liquid first-wall and the blanket. In this study, 9Cr2WVTa ferritic steel was chosen as a structural material because it has low activation, which can help to extend the lifetime of the reactor system. The main objective of this study is to investigate the neutron flux according to the mixture components, radial, energy spectrum in the designed hybrid reactor system for the selected fluids, library and structural material.

2 Method

2.1 Numerical Calculations and Neutron Flux

The radial structure of the hybrid reactor system is shown in Table 1. Calculation of all the parameters of the fission and fusion reactors, accelerator-driven systems, and other areas of nuclear technology depend on the cross-section data. The experimental data are limited for neutron-produced reactions in certain energy ranges. The data of the ENDF/B-VII libraries are very important for theoretical calculations [8-12]. This study was performed with neutron wall loadings of 10 MW/m^2 and fusion power of 4000 MW.

Table 1. The radial build of the hybrid reactor system design

Inboard side		Outboard side	
Zone	**r (cm)**	**Zone**	**r (cm)**
SS316LN	276	Plasma	667
Vacuum vessel[a]	278	SOL	695
SS316LN	294	Liquid First Wall[d]	697
GAP	296	Be[e]	700
Shield[b]	301	Blanket[d]	750
Ferritic Steel[c]	350	Ferritic Steel[c]	754
Blanket[d]	354	Shield[b]	804
Be[e]	404	GAP	838
Liquid First Wall[d]	407	SS316LN	840
SOL	409	Vacuum vessel[a]	866
Plasma	437	SS316LN	868

[a] 80% SS316LN, 20% H_2O
[b] 60% 9Cr2WVTa, 40% (99-95% $Li_{20}Sn_{80}$-1-5% RG-Pu, 99-95% $Li_{20}Sn_{80}$-1-5% RG-PuF_4, 99-95% $Li_{20}Sn_{80}$-1-5% RG-PuO_2)
[c] 100% 9Cr2WVTa
[d] 99-95% $Li_{20}Sn_{80}$-1-5% RG-Pu, 99-95% $Li_{20}Sn_{80}$-1-5% RG-PuF_4, 99-95% $Li_{20}Sn_{80}$-1-5% RG-PuO_2
[e] 100% Be

It is important for neutronic calculations to know the flux distribution of the neutrons in a nuclear fission reactor, fusion reactor, and hybrid reactor. The neutron flux is defined as the total path of thermal neutrons in a unit volume around point r in a second. The neutron flux expresses a scalar value of dimensions $m^{-2}s^{-1}$. In the nuclear reactors, fast neutrons are nearly exclusively produced in the fission reactions. In the nuclear reactors, the fast neutrons gradually slow due to collisions with the atoms of the moderator and the various nuclear processes taking place in the reactor [13]. As these

* Corresponding author: mehtap.gunay@inonu.edu.tr

formations are the functions of neutron energy, the neutron flux distribution should also be expressed as a function of neutron energy. The Boltzmann equation is a common way to calculate neutron flux in a reactor, which is given in below:

$$\frac{1}{v}\frac{\partial}{\partial t}\phi(r,\Omega,E,t) + \Omega.\nabla\phi(r,\Omega,E,t) + \Sigma_t(r,E,t)\phi(r,\Omega,E,t) = q(r,\Omega,E,t) \quad (1)$$

$\frac{1}{v}\frac{\partial}{\partial t}\phi(r,\Omega,E,t)$= Change of neutron flux in unit time,
$\Omega.\nabla\phi(r,\Omega,E,t)$ = Neutron loss because of convection,
$\Sigma_t(r,E,t)\phi(r,\Omega,E,t)$ = Neutron loss because of nuclear reactions.
Terms in equation (1) for $q(r,\Omega,E,t)$ can be defined as follow:

$$q(r,\Omega,E,t) = \int_0^\infty dE' \int_{4\pi} d\Omega \Sigma(r,E' \rightarrow E,\Omega' \rightarrow \Omega)\phi(r,\Omega',E',t) + S(r,\Omega,E,t) \quad (2)$$

$\int_0^\infty dE' \int_{4\pi} d\Omega \Sigma(r,E' \rightarrow E,\Omega' \rightarrow \Omega)\phi(r,\Omega',E',t)$= Contribution of neutrons on neutron flux due to scattering,
$S(r,\Omega,E,t)$= Contribution of neutron source independent on the neutron flux.

The transport equation is currently handled by the Monte Carlo method. The Monte Carlo method provides an instrument to solve the equations by simulating the neutron activities. In this study, neutron transport calculations of the neutron flux distributions were investigated using MCNPX-2.7.0 to solve the Boltzmann transport equation (1), (2) with Monte Carlo code.

3 Numerical Results

In this study, hybrid reactor system has designed by using 99–95% $Li_{20}Sn_{80}$ + 1-5% RG-Pu, 99–95% $Li_{20}Sn_{80}$ + 1-5% RG-PuF_4, and 99–95% $Li_{20}Sn_{80}$ + 1-5% RG-PuO_2 fluids in liquid first wall, blanket and shield zones, ENDF/B-VII.0 evaluated nuclear data library from 10^{-11} to 20 MeV and 9Cr2WVTa structural material.

Fig. 1 shows variation of the neutron fluxes in the reactor system versus the mixture components of 99–95% $Li_{20}Sn_{80}$ + 1-5% RG-Pu, 99–95% $Li_{20}Sn_{80}$ + 1-5% RG-PuF_4, and 99–95% $Li_{20}Sn_{80}$ + 1-5% RG-PuO_2 in the liquid first wall, blanket and shield zones. In Fig. 1 was observed that the neutron flux increases with an increase in the heavy metals content in the system for selected the fluids. Fig. 1 shows the create a peak around 5% of Pu, due to the fact that Pu has the highest plutonium density and fission reactions.

Fig. 2 shows the radial neutron flux distribution in the designed system for the various mixture compositions of molten salt and heavy metals. The variation of the neutron flux distribution in the inboard and outboard of the reactor system decreases nearly exponentially with distance from the reactor center (plasma) for selected the fluids.

Fig. 3 shows the average neutron flux distribution according to neutron energy spectrum (0-20 MeV) in the reactor system for the selected fluids. In Fig. 3, it was seen that the average neutron flux distribution decreases between 0-14 MeV, a peak around 15 MeV and decrease later in the designed system.

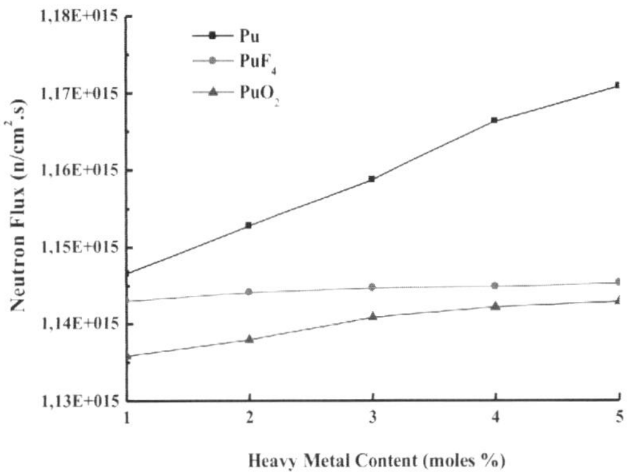

Fig. 1. The variation of the neutron fluxes in the reactor system versus the mixture components of molten salt and heavy metals.

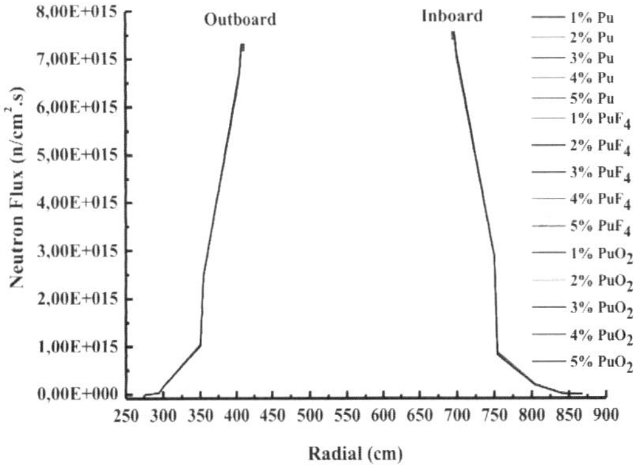

Fig. 2. The variation of the radial neutron flux distribution in the inboard and outboard of the reactor system for the selected fluids.

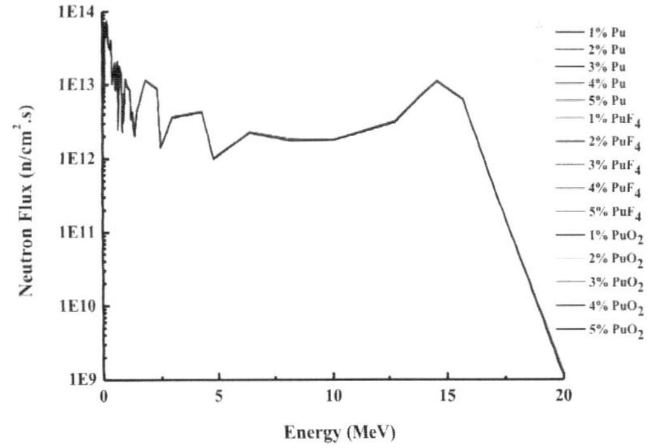

Fig. 3. The variation of the neutron flux distribution according to neutron energy spectrum in the reactor system for the selected fluids.

4 Conclusions

In this study were investigated the effect on the neutron flux distribution of selected fluids. The neutron flux increases with an increase in the heavy metals content for the selected fluids due to the fission reaction of plutonium isotopes, inelastic scattering of ^{7}Li and the exothermic $^{6}Li(n,\alpha)T$ reaction of ^{6}Li in the $Li_{20}Sn_{80}$ molten salt, (n,2n) reaction of beryllium, elastic and other nuclear reactions of atoms in the fuel zone. Therefore, in the calculation was observed that the radial neutron flux distribution decreases nearly exponentially with distance from the plasma for selected the fluids in the inboard and outboard of the reactor system. In this study, the average neutron flux distribution was investigated according to the neutron energy spectrum (0-20 MeV) in the system for the selected fluids. The neutron spectrum produces high energy γ-rays in the range of 5-10 MeV by (n,γ) reactions, 2 MeV fission neutrons and 14.6 MeV fusion neutrons. In the calculation was seen that the average neutron flux distribution due to fission, fusion and other nuclear reactions, and had a peak approximately 14-16 MeV, decreasing thereafter. The values between the 0-20 MeV energy gap were approximately the same values of the neutron flux distribution according to the energy in the system for ratios and types different fifteen of fluids.

In conclusion, the neutronic calculations obtained in the liquid first wall, blanket and shield zones with 95% $Li_{20}Sn_{80}$-5% RG-Pu fluid showed the best performance of the fifteen tested fluids.

Acknowledgements

This work was supported by İnönü University Research fund with the project number 2015/68.

References

1. S. Şahin, M. Übeyli, Energy Convers. Manage., **46**, 3185 (2005)
2. H.M. Şahin, Ann. of Nucl. Energy, **34**, 861 (2007)
3. B. Şarer, et al., Fusion Sci. Technol., **52**, 107 (2007)
4. M. Günay, et al., Ann. of Nucl. Energy, **38(12)**, 2757 (2011)
5. M. Günay, et al., Ann. of Nucl. Energy, **55**, 292 (2013)
6. IAEA, International Atomic Energy Agency, IAEA- TECDOC-1349 (2003)
7. M. Piera, et al., Energy Convers. Manage., **51**, 1758, (2010)
8. A. Ying, et al., APEX Interim Report, (1999)
9. B. Şarer, et al., Ann. of Nucl. Energy, **36**, 417 (2009)
10. M. Günay, et al., Ann. of Nucl. Energy, **53**, 59 (2013)
11. M.B. Chadwick, et al., Nucl. Data Sheets, **107**, 2931, (2006)
12. D.B. Pelowitz, MCNPX User's Manual, Version 2.7.0, LA-CP-11-00438 (2011)
13. J.J. Duderstadt, G.A. Moses, New York: John Wiley & Sons, (1982)

 DOI: 10.1051/epjconf/201715401026

Three-dimensional Monte Carlo calculation of some nuclear parameters

Mehtap Günay[*] , *Gökmen* Şeker

İnönü University, Science and Art Faculty, Physics Department, Malatya, Turkey

Abstract. In this study, a fusion-fission hybrid reactor system was designed by using 9Cr2WVTa Ferritic steel structural material and the molten salt-heavy metal mixtures 99–95% $Li_{20}Sn_{80}$ + 1-5% RG-Pu, 99–95% $Li_{20}Sn_{80}$ + 1-5% RG-PuF_4, and 99–95% $Li_{20}Sn_{80}$ + 1-5% RG-PuO_2, as fluids. The fluids were used in the liquid first wall, blanket and shield zones of a fusion–fission hybrid reactor system. Beryllium (Be) zone with the width of 3 cm was used for the neutron multiplication between the liquid first wall and blanket. This study analyzes the nuclear parameters such as tritium breeding ratio (*TBR*), energy multiplication factor (*M*), heat deposition rate, fission reaction rate in liquid first wall, blanket and shield zones and investigates effects of reactor grade Pu content in the designed system on these nuclear parameters. Three-dimensional analyses were performed by using the Monte Carlo code MCNPX-2.7.0 and nuclear data library ENDF/B-VII.0.

1 Introduction

The fuels used in hybrid reactor systems are generally deuterium–tritium (D-T) or deuterium–deuterium (D-D) fuels. When D-T fuel enters the fusion reaction, 14.1 MeV fusion neutrons and 3.5 MeV alpha particles are released. The plasma is surrounded by a wall made of fertile material, which cannot undergo the fusion reaction with thermal neutrons, but can undergo conversion with high energy neutrons, such as the 14.1 MeV neutrons produced by the fusion reaction. Thus, the high energy fusion neutrons convert fertile materials to fissile fuel, fission neutrons, and produce energy. The hybrid reactor produces 30 times more nuclear fuel per nuclear energy quantity than fast reactors [1-6].

In this study, a hybrid reactor system was designed. The hybrid reactor system can generate secure energy in large quantities with D-T fuel usage and subcritical study. Furthermore, it enables the production of a self-sufficient fuel for the reactor through the reaction of the neutrons released by the plasma. Reactor-grade (RG) plutonium is found in spent nuclear fuel. Isotopic distribution of RG-Pu has composition 1% ^{238}Pu, 62% ^{239}Pu, 24% ^{240}Pu, 8% ^{241}Pu and 5% ^{242}Pu [7]. The reactor-grade Pu in present nuclear power stations as fuel can use to reduce the trans-uranium content of spent nuclear fuel. in this study, radioactive materials with a plutonium additive were used in the hybrid reactor system to reduce the amount of reactor grade Pu used.

2 Method

2.1 Numerical Calculations

Hybrid reactor system used in the study is in the shape of a torus. The radial structure of designed the hybrid reactor system is shown in Table 1. Calculation of all the parameters of the fission and fusion reactors and other areas of nuclear technology depend on the cross-section data. The data of the ENDF/B-VII libraries are very important for theoretical calculations [8-10]. Analysis was performed for neutron wall loading 10 MW/m^2 and fusion power 4000 MW.

Table 1. The radial build of the hybrid reactor system design

Inboard side		Outboard side	
Zone	r (cm)	Zone	r (cm)
SS316LN	276	Plasma	667
Vacuum vessel[a]	278	SOL	695
SS316LN	294	Liquid First Wall[d]	697
GAP	296	Be[e]	700
Shield[b]	301	Blanket[d]	750
Ferritic Steel[c]	350	Ferritic Steel[c]	754
Blanket[d]	354	Shield[b]	804
Be[e]	404	GAP	838
Liquid First Wall[d]	407	SS316LN	840
SOL	409	Vacuum vessel[a]	866
Plasma	437	SS316LN	868

[a] 80% SS316LN, 20% H_2O
[b] 60% 9Cr2WVTa, 40% (99-95% $Li_{20}Sn_{80}$-1-5% RG-Pu, 99-95% $Li_{20}Sn_{80}$-1-5% RG-PuF_4, 99-95% $Li_{20}Sn_{80}$-1-5% RG-PuO_2)
[c] 100% 9Cr2WVTa
[d] 99-95% $Li_{20}Sn_{80}$-1-5% RG-Pu, 99-95% $Li_{20}Sn_{80}$-1-5% RG-PuF_4, 99-95% $Li_{20}Sn_{80}$-1-5% RG-PuO_2
[e] 100% Be

2.2 Tritium Breeding

The tritium breeding ratio (*TBR*) is defined as the ratio of the rate of tritium production in the system to the rate of tritium burning in the plasma. The hybrid reactor has to

[*] Corresponding author: mehtap.gunay@inonu.edu.tr

breed the tritium it requires [11,12]. Tritium is bred through the reaction of lithium with neutrons. The $^6Li(n,\alpha)T$ reaction is produced with thermal neutrons, and the $^7Li(n,\alpha n')T$ reaction with fast neutrons. In the hybrid reactor system, the working liquid must be a lithium-containing medium to provide adequate tritium so that the plasma is self-sustaining and the fusion is a renewable energy source. the required *TBR* for a fission reactor conceptual design must be 1.1 [11-15]. *TBR* can be given as follows:

$$TBR = T_6 + T_7 \quad (1)$$

$$T_6 = \iint \Phi \,.\, \Sigma_{(n,\alpha)T}\, dE\, dV, \text{ for } ^6Li \quad (2)$$

$$T_7 = \iint \Phi \,.\, \Sigma_{(n,n'\alpha)T}\, dE\, dV, \text{ for } ^7Li. \quad (3)$$

In this study, $Li_{20}Sn_{80}$ molten salt was used, which has low melting temperatures and low vapor pressure, to acquire sufficient tritium breeding.

2.3 Energy Multiplication Factor, Heat Deposition Rate and Fission Reaction Rate

The energy multiplication factor *M* is the ratio of nuclear heating and source neutron energy incident on the first wall [16]. The energy produced must exceed the energy produced by the plasma. To obtain high thermal power, the *M* value should be as large as possible. It is desirable that *M* is greater than 1.2 in a D-T fueled reactor [17]. *M* can be defined as follows:

$$M = [(200x(\Phi \,.\, \Sigma_f) + 4.784xT_6 - 2.467xT_7)/14.1] + 1 \quad (4)$$

$$(\Phi \,.\, \Sigma_f) = \iint \Phi \,.\, \Sigma_f\, dE\, dV, \text{ total fission rate.} \quad (5)$$

The contributions of the neutron flux, fission, and other reactions to the integrated *M* and heat deposition rate are very important. The main contributions to the integrated *M* and heat deposition rate are from the neutron flux, the fission reaction of plutonium isotopes, and the exothermic $^6Li(n,\alpha)T$ reaction of $Li_{20}Sn_{80}$ molten salt.

3 Numerical Results

In this study, the fluids were composed of molten salt $Li_{20}Sn_{80}$ as the main constituent with increased mole fractions of heavy metals, 1-5% RG-Pu, RG-PuF_4, RG-PuO_2. The fluids used in order to decrease amount of reactor grade Pu were used in the liquid first wall, blanket and shield zones of hybrid reactor system.

Fig. 1 shows the *TBR* in the liquid first wall, blanket and shield regions per fusion source neutron for the various mixture compositions of molten salt and heavy metals. The increasing heavy metals contents means decreasing the Li that produces tritium. In Fig. 1, the *TBR* decreases due to decreasing the Li that produces tritium with an increase in the heavy metals contents.

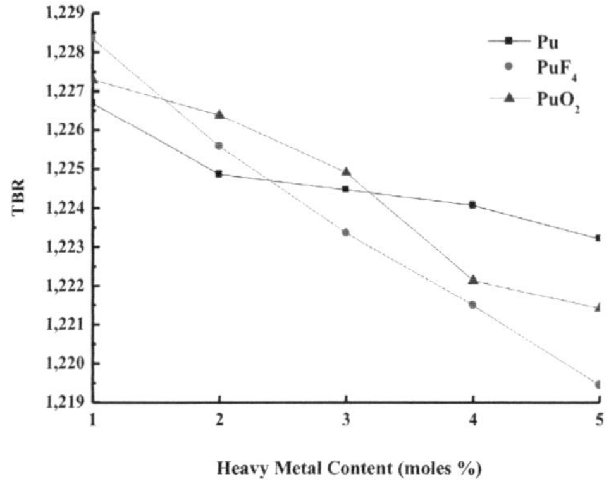

Fig. 1. The variation of the *TBR* value versus the mixture components of $Li_{20}Sn_{80}$, RG-Pu, RG-PuF_4, RG-PuO_2 in the liquid first wall, blanket and shield zones.

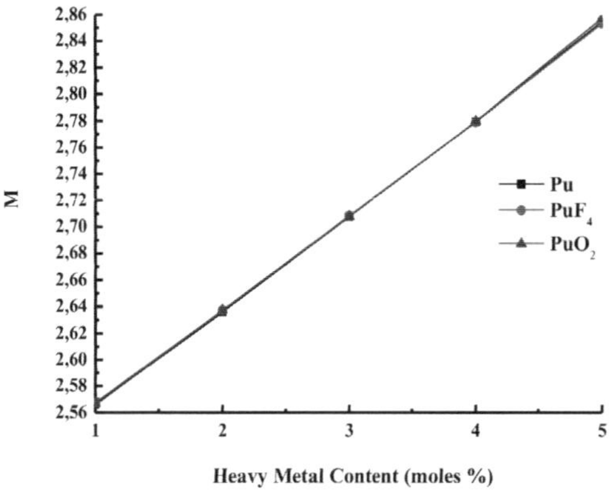

Fig. 2. The variation of the energy multiplication factor *M* versus the mixture components of $Li_{20}Sn_{80}$, RG-Pu, RG-PuF_4, RG-PuO_2 in the liquid first wall, blanket and shield zones.

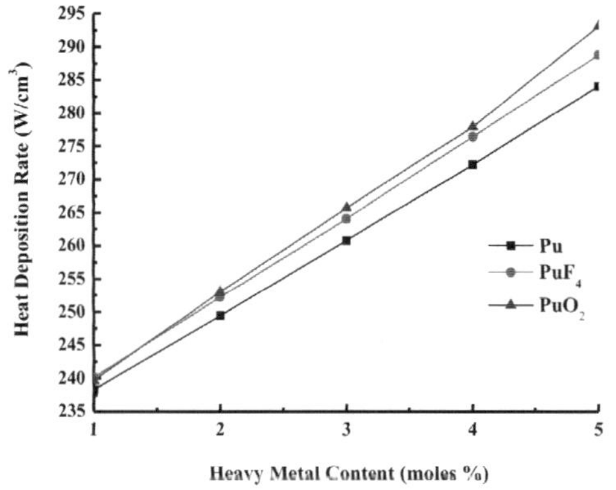

Fig. 3. The variation of the heat deposition rate versus the mixture components of $Li_{20}Sn_{80}$, RG-Pu, RG-PuF_4, RG-PuO_2 in the liquid first wall, blanket and shield zones.

Fig. 2 and Fig. 3 shows the energy multiplication factor *M* and the heat deposition rate in the liquid first wall, blanket and shield regions per fusion source neutron for the various mixture compositions of molten salt and heavy metals. In Fig. 2 and Fig. 3, it was found that *M* and the heat deposition rate increases linearly for increase of fission reaction rate with increasing of RG-Pu, RG-PuF_4, RG-PuO_2 heavy metals contents. The *M* value was obtained to be between 2.56 and 2.87 values. This value is greater than 1.2 value desirable of *M*.

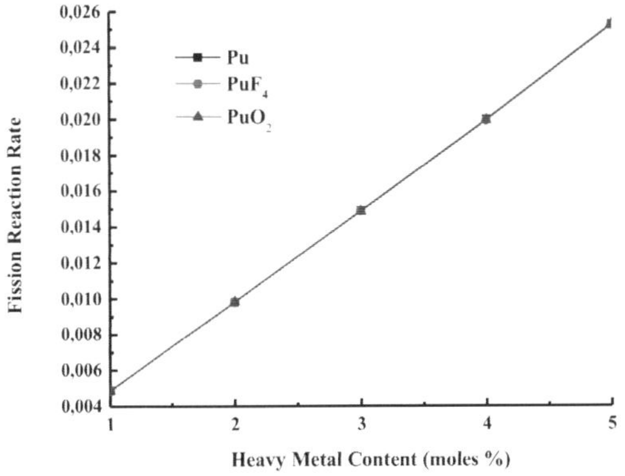

Fig. 4. The variation of the fission reaction rate versus the mixture components of $Li_{20}Sn_{80}$, RG-Pu, RG-PuF_4, RG-PuO_2 in the liquid first wall, blanket and shield zones.

Fig. 4 shows the variation of the fission reaction rate versus the mixture components of $Li_{20}Sn_{80}$, RG-Pu, RG-PuF_4, RG-PuO_2 in the liquid first wall, blanket and shield zones. Fig. 4 shows that the fission reaction rate increases linearly in the liquid first wall, blanket and shield regions with increasing of the heavy metals contents in mole percent.

4 Conclusions

In the study the effect of reactor grade Pu content in the designed system on three-dimensional neutronic measurements, such as *TBR*, *M*, heat deposition rate and fission reaction rate, was investigated.

In the calculation was observed that the *TBR* decreases with an increase in the heavy metals contents. The increasing heavy metals contents means decreasing the Li that produces tritium. This decrease between minimum and maximum rates of the heavy metals was obtained in approximately the same rate for RG-Pu, RG-PuF_4, RG-PuO_2. It was found that *TBR* values calculated with increase of the heavy metals contents higher than 1.1 the limit value. It was found that the greatest contribution to *TBR* comes from RG-PuF_4 in around 1% of the heavy metal content.

In the calculation, it was observed that the *M* and heat deposition rate increase with an increase in the heavy metal content for the selected fluids, owing to an increase of the fission reaction rate. All *M* values found from selected fluids are too close to each other and are approximately two times higher than 1.2 limit value of *M*. The increase between the minimum and maximum heat deposition rates and *M* values for the different heavy metals was approximately the same. It was found that the greatest contribution to *M* and the heat deposition rate comes with a fluid content of 95% $Li_{20}Sn_{80}$ + 5% RG-PuO_2.

In conclusion, it was found that reactor grade Pu amount decrease and neutronic calculations yield desirable values when one uses fluids of 95% $Li_{20}Sn_{80}$-5% RG-Pu, RG-PuF_4, RG-PuO_2, in comparison to the other ratios and types of fluids in the liquid first wall, blanket and shield zones.

Acknowledgements

This work was supported by İnönü University Research fund with the project number 2015/68.

References

1. S. Şahin, M. Übeyli, Energy Convers. Manage., **46**, 3185 (2005)
2. H.M. Şahin, Ann. of Nucl. Energy, **34**, 861 (2007)
3. B. Şarer, et al., Fusion Sci. Technol., **52**, 107 (2007)
4. M. Günay, et al., Ann. of Nucl. Energy, **38(12)**, 2757 (2011)
5. M. Günay, et al., Ann. of Nucl. Energy, **55**, 292 (2013)
6. M. Günay, et al., Ann. of Nucl. Energy, **63**, 157 (2013)
7. IAEA, International Atomic Energy Agency, IAEA- TECDOC-1349 (2003)
8. B. Şarer, et al., Ann. of Nucl. Energy, **36**, 417 (2009)
9. M. Günay, et al., Ann. of Nucl. Energy, **53**, 59 (2013)
10. M. Günay, et al., Ann. of Nucl. Energy, **63**, 432 (2014)
11. M. E. Sawan, et al., Fusion Engineering and Design, **81**, 1131 (2006)
12. J. Jung, et al., Nuclear Technol./Fusion, **4**, 361 (1983)
13. M. A. Abdou, et al., Fusion Technol., **9**, 250 (1986)
14. W. Kuan, et al., Fusion Technol., **35**, 309 (1999)
15. M. Z. Youssef, et al., Fusion Technol., **9**, 286 (1986)
16. S. Şahin, et al., Energy Convers. Manage., **45**, 1497 (2004)
17. A. H. Dhaba'an, et al., Progress in Nuclear Energy **29**, 1 (1995)

Investigation of spherical and cylindrical catural Iridium targets by photonuclear reaction

Mehmet Emin Korkmaz[1,*], *Osman* Agar[1], and *Mustafa* Yigit[2]

[1]Karamanoğlu Mehmetbey University, Faculty of Sciences, Department of Physics, Karaman, Turkey
[2]Aksaray University, Faculty of Arts and Sciences, Department of Physics, Aksaray, Turkey

Abstract. In this study, natural iridium consisting of Ir-191 and Ir-193 isotopes has been irradiated with 21 MeV photons. The distribution of photons, electrons and neutrons fluxes in the spherical and cylindrical natural iridium target have been calculated using MCNPX 2.7.0 Monte Carlo code. The intensity of the photon fluxes on both targets has been compared to the 10^6 particle story by showing them as mesh and optimizing the two targets.

1 Introduction

Nowadays particle beams are produced by tens of thousands of accelerators in various fields such as hospitals, manufacturing and printing plants, industrial laboratories, ports. More than 350 cyclotrons around world have been producing radioactive isotopes for use in medical applications, *e.g.* PET scan [1]. The radionuclides have been generally used at nuclear medicine for diagnostic purposes of some diseases as well as in therapy [2,3]. Recently, electron radiotherapy method has been of interest to carry out in the treatment of cancerous cells. The basic component of photonuclear radioisotope production by linear electron accelerators (LINACs) is a bremsstrahlung converter. A high Z photon converter which is made of tantalum (Ta) or tungsten (W) absorbs the high-energy electrons produced by LINAC. Thereby the energies of electrons are converted into bremsstrahlung photons [4]. On the other hand, nuclear data are needed in many applications which energy related investigations are fusion [5-8], fission, accelerator driven systems, etc. and also non-energy related investigations are medical radionuclide production [2,3], astrophysics, radiation therapy, etc [9]. Strong beta emitter Ir-192 ($T_{1/2}$=74 days) has been produced in the wire to use the inner radiotherapy source for prostate cancer treatment. Iridium which is one of the rare platinum group elements on the earth is used for various purposes in medicine and industrial since it's extremely resistant against corrosion and hot. Moreover, there are some application areas such as ignition plug, satellite systems, and industrial monitoring [10-12]. In general, the most stable radioisotope Ir-192 can occur as a consequence of either bombardment of Ir-191 with neutron in a nuclear reactor or separation of a neutron from Ir-193 by photonuclear reaction. In this study, natural iridium targets with two different geometries have irradiated with photon beam of 21 MeV energy produced from a linear accelerator by using Monte Carlo simulations.

2 Material and Method

2.1 Monte Carlo Method and Photonuclear reaction

Nuclear methods or models are widely needed to estimate nuclear data, especially if the measured data are not available or are unlikely to be produced because of the experimental difficulty [13,14]. Monte Carlo is a statistical method. This method is especially useful in complex problems that cannot be solved by deterministic methods [15]. Nowadays the best calculation tool in medical radiation research is the computer programs that use Monte Carlo (MC) methods. MC methods have been used in various radiation therapy branches such as radiation therapy devices, simulations of sources, dose calculations in targets with some geometry. The method is based on the use of random numbers and probability statistics in order to research any problems. Thus, it provides a reasonable approach at higher accuracy in the solution of problems according to the other techniques [16]. Otherwise radiotherapy application of MC method has a disadvantage since it consumes a lot of time. The calculation time depends on the number, energy and type of the produced particle. In order to obtain a good statistical result in calculations, millions of particles must be traced [17]. Therefore, it needs parallel-working super computers with high calculation power in some MC studies [18]. In this work, all calculations were applied by using MCNPX 2.7.0 code developed by Los Alamos National Laboratory (LANL) [19].

2.2 Iridium Target System

In this study, two different iridium targets with spherical and cylindrical geometries were irradiated photon beam of 21 MeV energy and 1 cm radius. It has been calculated for 10^6 particle histories to decrease the statistical errors in simulations of MCNP-X code. The calculations of MC simulation were completed using

* Corresponding author: mkorkmaz@kmu.edu.tr

DOI: 10.1051/epjconf/201715401027

Intel Xeon PC workstation of a double processor 2.4 GHz and 48 GB memory space. The natural iridium target seen in two dimensions is a sphere which has a radius of 1 cm and located at origin as given in Fig. 1a.

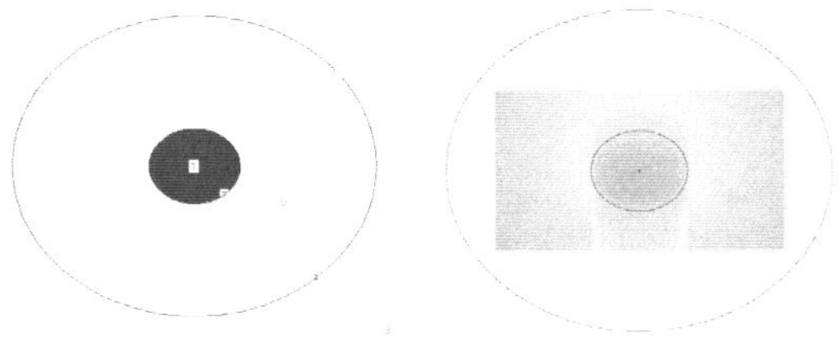

(a) (b)

Fig. 1. (a) The geometry of the spherical natural iridium target (b) The mesh calculations of the photon flux distributions at spherical target

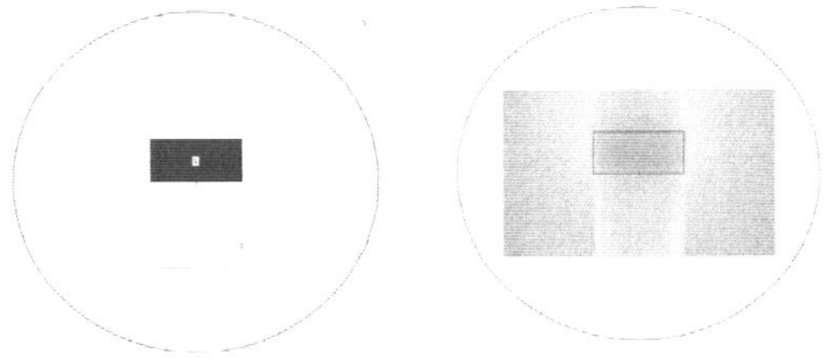

(a) (b)

Fig. 2. (a) The geometry of the cylindrical natural iridium target (b) The mesh calculations of the photon flux distributions at cylindrical target

Photon source is 2 cm away from the origin. Fig. 1b shows the scattering depending on arrival direction of photon beams for 180000 particle histories on the spherical target and mesh calculations of photon flux density. The photon flux density decreases outward from the center of target. Similarly, the iridium target in Fig. 2a is a cylindrical form with a radius of 1 cm and a height of 1 cm. Fig. 2b represents the mesh calculations of photon flux depending on arrival direction of photon beams for 21000 particle histories. The photon flux density in the center of target is higher than the other side.

3 Results and Discussion

3.1 Particle Flux

Photon, electron and neutron flux distributions of two different natural iridium targets were calculated and compared. The spherical and cylindrical targets were irradiated with photon beam of 21 MeV mono-energetic. Fig. 3 illustrates photon, electron and neutron flux distributions depending on energies for two targets. Although both targets have similar flux distributions, flux of cylindrical are higher than those of spherical target, especially photoneutron flux at 0-13 MeV.

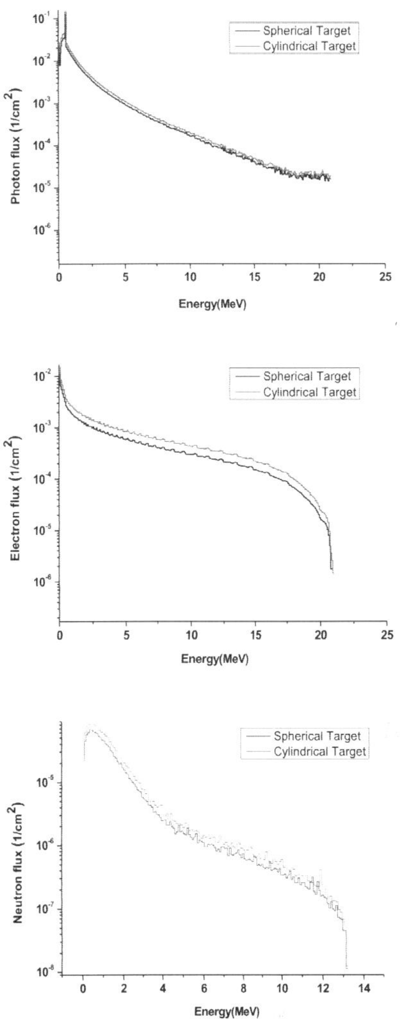

Fig. 3. Photon, electron and neutron flux distributions of spherical and cylindrical targets depending on energies.

3.2 Energy Deposition

Fig. 4 represents the stored energies of photon, electron and neutron resulting from nuclear reaction that occur at spherical and cylindrical targets. The energy storage of photons shows approximately similar for different target types as seen in Figure. From the other side, the energy depositions of electrons and neutrons at cylindrical target are quite different and higher than those of spherical.

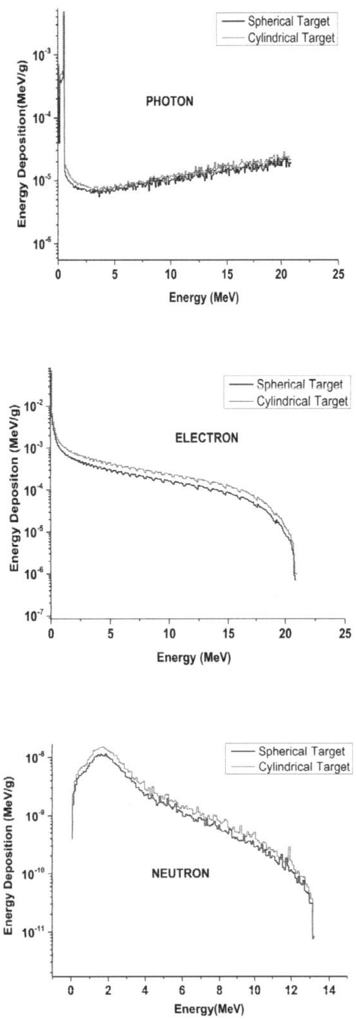

Fig. 4. The energy depositions of photons, electrons and neutrons at spherical and cylindrical targets depending on energies

4 Conclusions

The medical electron accelerators are powerful photon sources. Therefore, the photonuclear reaction data can be useful for different technological and scientific developments. One of the most important applications is to produce Ir-192 isotope resulting from photonuclear reaction of Ir-193 stable isotope. Especially, Ir-192 isotope is very important in the treatment of cancer. This isotope has potential to be produced by linear accelerator working at region of 5-25 MeV energy in hospital. Cylindrical wire-shaped Ir-192 isotope used in brachytherapy is, in general, produced by irradiated with neutrons in reactors. This work shows that spherical-shaped natural iridium target can be produced and give a similar result to those of cylindrical target. Production of medical radioisotopes will increase together with new accelerator technologies. Methods that required R&D such as beam transmission and targeting for ion therapy will highly be developed.

This study was supported by Karamanoglu Mehmetbey University Scientific Research Project (37-M-16).

References

1. U.S. Department of Energy Report, *Accelerators for America's Future*, (2010)
2. A. Kara, T. Korkut, M. Yiğit, E. Tel, Kerntechnik **80,** 270 (2015)
3. M. Yiğit and E. Tel, Ann. Nucl. Energy **69,** 44 (2014)
4. H. Ghiasi and A. Mesbahi, Iran J. Radiat. Res. **8,** 187 (2010)
5. E. Tel, S. Akca, A. Kara, M. Yiğit, A. Aydin, J. Fusion Energ. **32**, 531 (2013)
6. H. Şahan, E. Tel, M. Yiğit J. Fusion Energ. **34**, 16 (2015)
7. M. Yiğit, E. Tel and İ.H. Sarpün, Nucl. Instrum. Meth. A **385**, 59 (2016)
8. H. Korkut, T. Korkut, A. Kara, M. Yiğit, E. Tel, J. Fusion Energ. **35**, 591 (2016)
9. M. Yiğit, E. Tel, A. Kara, J. Fusion Energ. **32**, 362 (2013)
10. S.S. Belyshev, A.N. Ermankov, B.S. Ishkhanov, V.V. Khankin, A.S. Kurilik, A.A. Kuznetsov, V.I. Shvedunov, K.A. Stopani, Nucl. Instrum. Meth. A **745**, 133 (2014)
11. C. Beinke, A. Ben-Shenlomo, M. Abend, M. Port, Radiat. Res. **184**, 66 (2015)
12. I. Gerardy, J. Rodenas, S. Gallardo, Appl. Radiat. Isot. **69**, 1108 (2011)
13. E. Tel, M. Yiğit, G. Tanir, J. Fusion Energ. **31,**184 (2012)
14. E. Tel, M. Yigit, G. Tanir, J. Fusion Energ. **32,** 273 (2013)
15. M.E. Korkmaz, M. Yiğit, O. Agar, Kerntechnik **78**, 193 (2013)
16. K. Jabbari, A. Sarfehnia, E.B. Podgorsak, J.P. Seuntjens, Phys. Med. Bio. **52,** 1171 (2007)
17. M. Zoubair, T. El Bardouni, O. Allaoui, Y. Boulaich, B. El Bakkari, C. El Younoussi, H. Boukhal, E. Chakir, WJNST **3,** 14 (2013)
18. T. Liu, X.G. Xu, C.D. Carothers, Ann. Nucl. Energy **82,** 230 (2015)
19. D.B. Pelowitz, *MCNPX User's Manual*, (Version 2.7.0, LA-CP-11-00438, 2011)

EPJ Web of Conferences **154**, 01028 (2017) DOI: 10.1051/epjconf/201715401028
TESNAT 2017

The (α,n) reaction cross-section calculations for some isotopes of iron group elements in the range of 5-20 MeV

Ercan Yildiz*

Kırıkkale University, Department of Physics, Kırıkkale, Turkey

Abstract: Iron group elements (Sc, Mn, Fe, Co, Ni) are especially very important because structural material selection in design of fusion–fission reactors is very crucial. Obtained results from the nuclear reactions using structural materials can be used developing for these structural materials some isotopes. For this reason, in this study the cross sections of $^{45}Sc(\alpha,n)^{48}V$, $^{55}Mn(\alpha,n)^{58}Co$, $^{54}Fe(\alpha,n)^{57}Ni$, $^{59}Co(\alpha,n)^{62}Cu$, $^{62}Ni(\alpha,n)^{65}Zn$ reactions have been calculated at 5-20 MeV energy ranges by using TALYS 1.8 and NON-SMOKER codes. Obtained results were compared experimental data from EXFOR database.

1. Introduction

Nuclear fusion is a safe and clean energy source which serves unlimited energy to mankind. One of the most important advantages of a fusion energy system is the abundant fusion fuel availability in the nature, contrary to relatively scarce fission fuel resources. Furthermore, a fusion energy system has aspects of an attractive product with respect to safety and environmental advantages compared to other energy sources. Thus, market penetration of fusion energy reactors may create a revolution in energy generation in near future. Nevertheless, the market penetration of merchant pure fusion reactors have not been expected before the year 2050. On the other hand, the combination of a fusion and fission reactor may have realistic chances for a relatively earlier introduction of fusion power plants for electricity production [1].

The first wall, blanket system is the structural core materials for nuclear fusion reactors. The success of nuclear fusion reactors is highly dependent on performance and reliability of the first wall, blanket system which is the core structural materials for fusion reactors. Therefore, the selection of structural materials is extremely important for fusion reactor technology [2-5]. These materials can be classified their activation characteristics. Sc, Mn, and Fe are considered as low activation elements while elements such as Nickel (Ni), Cobalt (Co) are some of the high activation materials [6].

The knowledge of nuclear data in the fusion technology has a great importance in the domains of nuclear structure, nuclear decay, and nuclear reaction. The experimental measurements model theory calculations and systematic predictions are rather useful to provide this knowledge, especially for predicting the cross-sections of nuclear reaction [7].

In this study, cross sections of the $^{45}Sc(\alpha,n)^{48}V$, $^{55}Mn(\alpha,n)^{58}Co$, $^{54}Fe(\alpha,n)^{57}Ni$, $^{59}Co(\alpha,n)^{62}Cu$, $^{62}Ni(\alpha,n)^{65}Zn$ reactions have been calculated with the TALYS 1.8 [8] and NON-SMOKER [9] codes. Obtained results were compared with experimental data from EXFOR [10]

2. Results and Conclusions

The (α,n) reactions cross sections of given reactions were calculated in the incident alpha energy range of 5 to 20 MeV by using default values in the TALYS 1.8 and NON-SMOKER codes. We compared to the calculated data and experimental data from EXFOR database in Figs. 1-5.

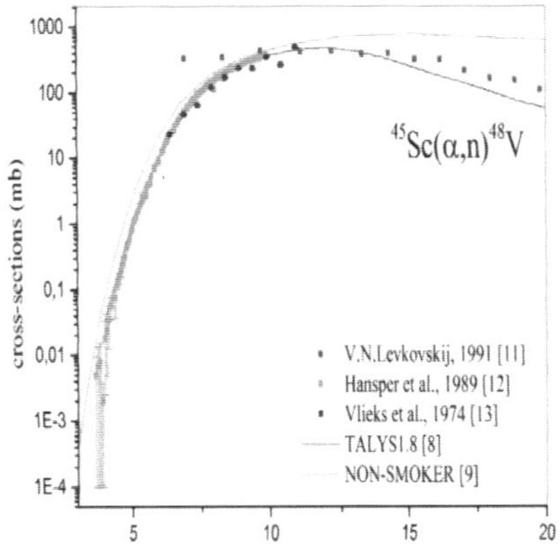

Fig. 1. Comparison of experimental and evaluated cross-sections of $^{45}Sc(\alpha,n)^{48}V$ as a function α energy.

*Corresponding Author: ercan7306@hotmail.com

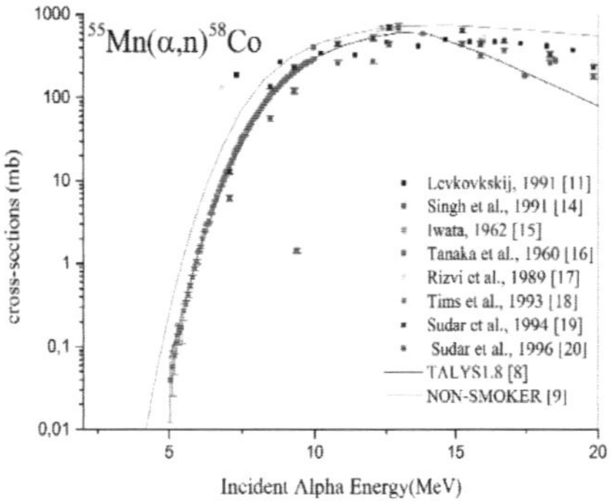

Fig. 2. Comparison of experimental and evaluated cross-sections of ^{55}Mn(α,n)^{58}Co as a function α energy.

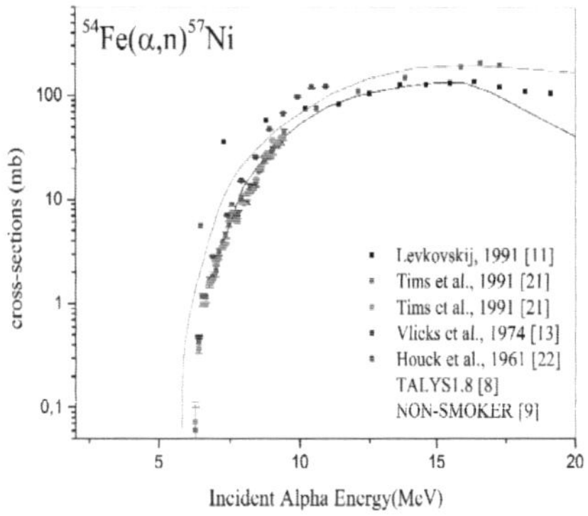

Fig. 3. Comparison of experimental and evaluated cross-sections of ^{54}Fe(α,n)^{57}Ni as a function α energy.

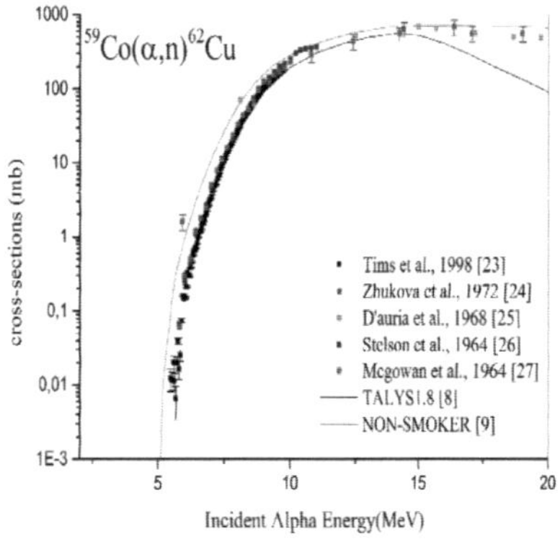

Fig. 4. Comparison of experimental and evaluated cross-sections of ^{59}Co(α,n)^{62}Cu as a function α energy.

We can see that the calculated data by TALYS 1.8 are agrement with all the experimental data for ^{45}Sc(α,n)^{48}V reaction.

It can be seen that the calculated cross-section data by TALYS 1.8 are agrement with the experimental data at low energies for ^{55}Mn(α,n)^{58}Co, ^{54}Fe(α,n)^{57}Ni, ^{59}Co(α,n)^{62}Cu, ^{62}Ni(α,n)^{65}Zn reaction at low energies especially.

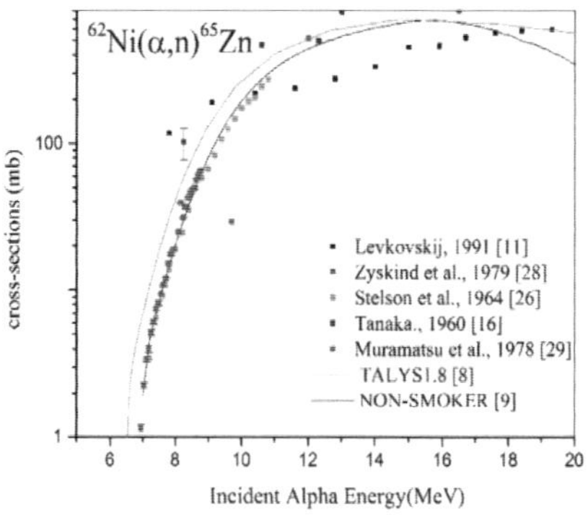

Fig. 5. Comparison of experimental and evaluated cross-sections of ^{62}Ni(α,n)^{65}Zn as a function α energy.

References

[1] M. Ubeyli, Materials and Design **28** (2007)
[2] W.R. Meier et al., Lawrence livermore national laboratory,LLNL-JRNL-416976 (2009)
[3] IAEA Publication, Development of radiation resistant reactor core structural materials,http://www.iaea.org/About/Policy/GC/GC51/GC51InfDocuments/English/gc51inf-3-att7_en.pdf
[4] M. Rubel, Trans Fusion Sci Technol **53** (2008)
[5] M. Victoria et al., Nucl. Fusion **41**, 8 (2001)
[6] P.M. Raole et al., Trans. IIM **62** (2009)
[7] E. Tel et al., J. Fusion Energy **35,** (2016)
[8] A. J. Koning, H. Hilaire, S. Goriely, TALYS-1.8, NRG, Netherland, http://www.talys.eu (2017)
[9] http://nucastro.org/nonsmoker.html
[10] https://www-nds.iaea.org/exfor/exfor.htm
[11] V.N. Levkovskij, *Act.Cs.by Moscow* (1991)
[12] V.Y.Hansper et al., Nuclear Physics,**551** (1993)
[13] A.E.Vlieks et al., Nuclear Physics, **224** (1974)
[14] B.P. Singh et al., Canadian Journal of Physics Vol.**69** (1991)
[15] S.J. Iwata, Journal of the Physical Society of Japan Vol.**17** (1962)
[16] S.Tanaka, Journal of the Physical Society of Japan Vol.**15** (1960)
[17] I.A.Rizvi, et al., Canadian Journal of Physics Vol.**67** (1989)
[18] S.G. Tims, Nuclear Physics, Section A Vol.**563** (1993)
[19] S .Sudar, et al., Phys. Rev, Part C, Nuclear Physics Vol.**50** (1994)
[20] S .Sudar, et al., Phys. Rev, Part C, Nuclear Physics Vol.**53** (1996)
[21] S.G. Tims, Nuclear Physics, Section A Vol.**524** (1991)
[22] F.S.Houck, et al., Phys. Rev Vol.**123** (1961)
[23] S.G. Tims, Nuclear Physics, Section A Vol.**483** (1998)
[24] O.A. Zhukova, et al.,Fizika Vol.**16** (1972)
[25] J.M. D'auria, et al., Phys. Rev Vol.**168** (1968)
[26] P.H. Stelson, et al.,Phys. Rev.**133** (1964)
[27] F.K. Mcgowan, et al., Phys. Rev.**133** (1964)
[28] J.L. Zyskind, et al., Nuclear Physics, Section A Vol.**331** (1979)
[29] H. Muramatsu, et al., Applied Radiation and Isotopes Vol.**29** (1978)

EPJ Web of Conferences **154**, 01029 (2017) DOI: 10.1051/epjconf/201715401029
TESNAT 2017

Multilinear analysis of Time-Resolved Laser-Induced Spectra of U(VI) containing natural water samples

Jakub Višňák [1,2,3,*]

[1]*Dep. of Nuclear Chemistry, FNSPE, Czech Technical University, Břehová 7, Prague 1, Czech Rep.*
[2]*Dep. of Chemical Physics and Optics, Faculty of Mathematics and Physics, Charles University, Prague 2, Czech Rep.*
[3]*J. Heyrovský Institute of Physical Chemistry, Dolejškova 2155/3, 182 23 Prague 8, Czech Rep.*

Abstract. Natural waters uranium level monitoring is of great importance for health and environmental protection. One of possible detection method is the Time-Resolved Laser-Induced Fluorescence Spectroscopy (TRLFS), which offers possibility to distinguish different uranium species (that is important since they have different sorption properties, mobility and requires different protection measures). Samples originate from Saxony, Germany and have been provided by Wismut GmbH company. They have been characterized by total elemental concentrations and TRLFS spectra. Uranium in the samples is supposed to be in form of uranyl(VI) complexes mostly with carbonate (CO_3^{2-} and bicarbonate HCO_3^-) and to lesser extend with sulphate (SO_4^{2-}), arsenate (AsO_4^{3-}), hydroxo (OH^-), nitrate (NO_3^-) and other ligands. Presence of alkaline earth metal dictations (M= Ca^{2+}, Mg^{2+}, Sr^{2+}) will cause most of uranyl to prefer ternary complex species, e.g. $M_n(UO_2)(CO_3)_3^{2n-4}$ ($n \in \{1;2\}$). From species quenching the luminescence, Cl^- and Fe^{2+} should be mentioned. Measurement has been done under cryogenic conditions to increase luminescence signal. Preliminary experimental study on cryogenic freezing influence on sample speciation is presented for a well-defined UO_2^{2+} - SO_4^{2-} system (room temp. TRLFS vs. cryo TRLFS). Data analysis has been based on Singular Value Decomposition and monoexponential fit of corresponding loadings (for separate TRLFS spectra, the "Factor analysis of Time Series" (FATS) method) and Parallel Factor Analysis (PARAFAC, all data analysed simultaneously). From individual component spectra, excitation energies T_{00}, uranyl symmetric mode vibrational frequencies ω_{gs} and excitation driven U-Oyl bond elongation ΔR have been determined and compared with quasirelativistic (TD)DFT/B3LYP theoretical predictions to cross-check experimental data interpretation.

1 Motivation

This contribution presents a first step in longer run of both experimental analysis and theoretical chemical (quantum chemistry and molecular dynamics) studies of uranium speciation in natural water samples and subsequent study of possible chemical/physical remediation meeting criteria for health and environment protection.

A preliminary analysis of six sample TRLFS spectra (S1-S5,S9) by FATS method and of seven samples (the previously mentioned and S10) together by PARAFAC will be presented. The sample S6-S8 TRLFS spectra had too low signal:noise (s/n) ratio (due to luminescence quenching by Cl^-, Fe^{2+}, Mn^{2+} and organic compounds) and author wouldn't be able to present meaningful interpretation of analysis results (this can change in future when more measurement for similar samples and/or titration by reagent bonding the quencher will be done).

The speciation, i.e. the information on how is given total analytical concentration of uranium partitioned between different chemical forms (coexisting in the same sample in chemical equilibrium) is of environmental importance because different chemical species will have different physical, chemical and biological properties such as mobility, toxicity and will require different measures for remediation. For example, here studied samples are in pH range 5.5-9.3 and are aerobic (E_H range from 130-450 mV), are supposed (based on analysis presented below in contribution text) to consist dominantly of neutral charged ternary complex $Ca_2UO_2(CO_3)_3^0$, two-fold negatively charged $CaUO_2(CO_3)_3^{2-}$, $MgUO_2(CO_3)_3^{2-}$ (and to a lesser extend $UO_2(CO_3)_2^{2-}$, $UO_2(SO_4)_2^{2-}$) and to a lesser extend highly negative charged $UO_2(CO_3)_3^{4-}$. On of possibilities for remediation is interacting water in question with anion-exchange resins, but stable and by concentration dominating $Ca_2UO_2(CO_3)_3^0$ complex pose problem here, due to its electrical neutrality (and chemical equilibration

* Corresponding author: jvisnak@gmail.com

changed by resin might not be fast enough in given experimental setup). But not all physical-chemical properties should be reduced to electric charge only, of course.

2 Theoretical background

2.1 Uranyl compounds spectra and optimal spectroscopic parameter choice

The aqueous uranyl complex compound luminescence theoretical background is briefly discussed in [1]. To cut the basis it into single sentence–the luminescence corresponds to (by a good working hypothesis) single electronic transition on frontier molecular orbitals of central uranyl group (chromophore), is resolved by a symmetric stretching mode of UO_2^{2+} group and ligands can be seen as merely changing the excitation energy T_{00} of the transition, the vibrational frequencies ω_{gs}, ω_{es} and ΔR parameter mentioned later in text.

For practical reasons, it should be stressed that while standard literature information extracted from luminescence spectra (of individual chemical species – aqueous samples, almost in all cases, consists of several different species coexisting in chemical equilibrium) consists of mere three to six luminescence band positions in nm (given usually with 1 nm precision), author see this praxis as rather unfortunate.

The reason is that since uranyl compounds luminescence spectra are very similar to each other, one need to be very careful with spectroscopic parameter set to choose. It is easy to observe that band positions, however in to-energy-proportional unit (e.g. cm^{-1}), form by (two) parts linear function of their ordinary number (please see Fig. 1,2 and Fig. 5 in [1]). That is, both cold-bands are equidistant in cm^{-1} scale and hot-bands are equidistant in cm^{-1} scale. The respective slopes in linear dependence on ordinary peak number or "peak maxima spreads correspond to "effective symmetric stretching mode vibration frequency of uranyl, UO_2^{2+} (central) group" for electronic ground (cold-bands, ω_{gs}) and luminescence-active excited state (hot-bands, ω_{es}). The most energetic cold-band, the 0'→0 peak energy complete the three-parameter set (T_{00}, ω_{gs}, ω_{es}) describing all peak positions (no matter if they are three as well, or up to seven). While the linearity (or equidistanceness) of peak position is usually very strong (uranyl group anharmonicity $\omega_e x_e$ is below 15 cm^{-1} as compared to ω_{gs} = 870 cm^{-1} for pentaaqua complex in water under ambient conditions) it is still better to obtain T_{00} as an intercept in linear regression of peak cm^{-1} positions instead of just 0'→0 peak position only. But storing seven peak positions seems to be rather redundant.

On the other side, the luminescence spectra shouldn't be reduced to band position information only – the way signal is partitioned between different peaks (i.e. ratio of peak heights/areas under the peaks) provides and important information too. And since all spectroscopic parameters of individual uranyl chemical species coexisting in the same aqueous sample usually differ by quantity on and edge of experimental uncertainty (or even below) every single piece of non-redundant spectral information matter greatly. This information can be characterized again by a property with a direct quantum chemical meaning (and therefore accessible by an ab initio theoretical modelling) – the "excitational elongation" $\Delta R[pm]=R_{es}-R_{gs}$, meaning an absolute value of difference between U-Oyl equilibrium bond length in electronic excited state (R_{es}) and in electronic ground state (R_{gs}, for further information, please see [1], the one-parameter fit with linear harmonic oscillator Fracnck-Condon factors for pentaaqua uranyl is given in Fig. 6 there).

Another information might be provided by individual peak FWHMs and their shapes (possible asymmetry or deviation from gaussian/voigt shape), but since this information is much more measurement-setup-dependent (e.g. the aperture slit widening will cause peak widening) and much less easy to interpret, it has less sense to collect it.

For consistency check it is also important that certain independent spectroscopical measurements (different from TRLFS) can be used to determine the above-mentioned parameters–UV-VIS (T_{00} and ω_{es}, spectrophotometric measurements are possible even for sub milimolar to micromolar total uranium concentration range when light absorbance is measured in a very long capillary (as is practised at HZDR) and Raman (ω_{gs}), Excited state EXAFS (ΔR – from R_{es} if R_{gs} is measured by normal EXAFS). The IR spectroscopy would provide information on anti-symmetric stretching mode of uranyl central group (IR is possible only under special circumstances for aquesous samples, of course).

Spectroscopic property derived from temporal domain is the luminescence lifetime, τ_m (under cryogenic conditions is in the ms range, unlike the μs range corresponding to the ambient conditions). Each species is characterized by a single τ_m value (m is the index of chemical species) parametrizing the simplest mono-exponential luminescence decay model. However, the tm value even for a fixed species may vary sample to sample because of different concentration of quenchers and/or different major species chemical composition (pleas, see eq. (21) and (22) in [1]). This is addressed as "matrix effect" and can, to same extend, affect spectroscopical parameters derived from (emission) wave-length/wave-number domain (T_{00}, ω_{gs}, ω_{es}, ΔR, FWHMs) as well. The luminescence life-times are also dependent on temperature (approximately by an Arhenius Law for k_q parameters in eq. (22) of [1]) and phase (different in amorphous ice and liquid water even for the same temperature).

Interestingly, measureable changes in T_{00} and ω_{gs} of the same individual chemical components (UO_2^{2+}, UO_2SO_4, $UO_2(SO_4)_2^{2-}$ and $UO_2(SO_4)_3^{4-}$) have been detected between ambient and cryogenic conditions for uranyl – sulfate system (which has been measured under both conditions in one experimental campaign by author recently (the results will be published in near future) at HZDR []). This phenomenon has been well known to other experimenters at HZDR as well. Unfortunately, such a comparison, is not possible for uranyl – carbonate system since uranyl carbonates yield insufficient luminescence under ambient conditions. However, some of the experience learned on the uranyl – sulfate system "ambient vs. cryo" comparison will help to answer questions such as "Is the speciation (un)changed in the process of cryogenic cooling of the sample?". The general hope is that change is either small or predictable (and therefore, by thermodynamics based calculation correctable) and I will address this topic in my future contributions.

2.2 Multilinear experimental data analysis methods used – FATS and PARAFAC

2.2.1 Problem formulation

Since uranium total concentration in all studied samples is well below 0.003 mol.dm^{-3} (more concentrated solutions may exhibit self-absorption and luminescence signal might not be linear with respect to individual component concentrations), laser pulse energy has been around 500 μJ only and MCP chosen so measurement has been done inside the linear part of dynamic range of ICCD detector, we can write measured luminescence signal in in-th spectrum, $Y_i(\lambda)$, as a linear combination of (yet unknown) TRLFS spectra from individual chemical species,

$$Y_i(\lambda) = \sum_{m=1}^{b} C_{im} Z_m(\lambda) + n_i(\lambda), \qquad (3)$$

where λ is wavelength, C_{im} luminescence amount in i-th spectrum corresponding to m-th species (individual component) and Z_m spectrum of the individual chemical species (e.g. UO_2^{2+}, UO_2SO_4, UO_2CO3, $Ca_2UO_2(CO_3)_3^0$, $CaUO_2(CO_3)_3^{2-}$, $MgUO_2(CO_3)_3^{2-}$, ..., m is a positive integer corresponding to some of the above written species). The total number of distinguishable components is denoted as b. The n_i (λ) is the noise function. The spectrum index i can either represent given i-th delay t_i between laser excitation of a sample and ICCD camera luminescence signal collection (case of kinetic/time series (TS) as in FATS) or given i-th sample (when delay is kept constant for all samples), or in the most robust procedure, there can be mapping between index i values and doubles (t, k), where t represents delay and k sample number (all kinetic series analyzed together).

The most general multilinear fitting procedure based on Singular Value Decomposition (SVD) can be formulated as follows: let us assume that from SVD decomposition of measurement data matrix Y_{li} (spectral index $i \in \{1,2, \ldots, s\}$, wave-length index $l \in \{1, 2, \ldots, N\}$),

$$Y_{li} = \sum_j U_{lj} W_{jj} V_{ij}, \qquad (4)$$

or in matrix form

$$Y = U W V^T, \qquad (5)$$

we take first f components ($j \in \{1,2,\ldots,f\}$) as representing signal and filter out the remaining components ($j > f$).

Matrix U in (5) has orthonormal columns (and same dimensions as the original data matrix Y, the columns of U are called "subspectra", the first f of them represent orthonormal basis of subspace of R^N corresponding to signal spectra), matrix W is diagonal positive semi-definite with diagonal elements ("singular values") sorted from the greatest to the smallest and V columns are called "loadings", they form an orthonormal set and j-th loading elements V_{ij} represent relative amnout of j-th sub spectrum in i-th original spectrum. Again, we can think about columns of V (loadings) as basis vectors (in R^s, space of different concentration/amount profiles of individual chemical species). Filtering out noise-like small components mentioned in previous paragraph means that in later derivation only first f columns are considered for matrices U and V and that only first f x f diagonal block of W is considered as well.

Setting (5) equal to matrix variant of (3) (which reads

$$Y = Z C^T + E, \qquad (6)$$

) and using Ansatz

$$Z = U R^T, \qquad (7)$$

will lead to matrix equation

$$V = C R W^{-1}, \qquad (8)$$

entering fitting procedure described in detail later. The f x f square matrix R has yet unknown elements (which will be retrieved by the fit of (8)) and represent transition matrix between orthonormal basis of subspectra and nonortogonal set of (yet unknown) individual component spectra.

In general, we can consider gaussian likeli-hood functions for loadings V_{ij} (assumed to be statistically independent and random gaussian distributed around modelled average) variables (8), leading to problem of minimization of the objective function

$$\chi^2_{MEM}(R,\beta,\gamma) = \sum_{i,j} \frac{\left(V_{ij} - C_{im}^{(\text{model})}(\beta) R_{mj} W_{jj}^{-1}\right)^2}{\sigma_{ij}^2(\gamma)} + \sum_{i,j} \ln\left(\sigma_{ij}^2(\gamma)\right), \qquad (9)$$

where summation is over $i \in \{1,2, \ldots, s\}$ and $j \in \{1,2, \ldots, f\}$, σ_{ij}^2 stands for variance of V_{ij} and γ for a set of parameters of dependence σ_{ij}^2 on V_{ij} (for Poisson-like distribution model, we can consider

$$\sigma_{ij}^2(\gamma_i) = \gamma_i \cdot \underset{l}{\text{mean}} |Y_{li}|, \qquad (10)$$

) and $C_{im}^{(\text{model})}$ stands for model of luminescence amount „profile" in studied spectra.

2.2.2 SVD-based methods and models

Three different cases should be considered:

1) FATS analysis (The s spectra represent one TRLFS kinetic series measurement of one separate sample, to each i correspond given delay t_i), then $C_{im}^{(\text{model})}$ corresponds to the model of luminescence decay and for the simplest one, monoexponential, the parameters β are labeled τ_m and correspond to

$$C_{im}^{(\text{model})} = \Omega_m \exp(-\tau_m / t_i), \qquad (11)$$

where Ω_m is a prefactor corresponding to non-zero width of ICCD detector integration window (Δt, integration time). For single FATS analysis, Ω_m (12) can be omitted.

$$\Omega_m = \tau_m \left(1-\exp\left(-\Delta t / \tau_m\right)\right). \quad (12)$$

2) FACSC (Factor analysis connected to speciation computation) – in this case the s spectra correspond to s different samples measured within one fixed delay and $C_{im}^{(\text{model})}$ represent speciation model (i.e. (implicit) formula for m-th species molar concentration in i-th sample (i can be linked to i-th total ligand concentration $c_{L,i}$ or similar variable)). The parameters in this model can be, e.g., speciation/chemical equilibrium constants characterizing stability of individual components.

$$C_{im}^{(\text{model})} = C_m^{(speciation)}\left(\beta_m^{spec}; c_{L,i}\right), \quad (13)$$

3) FATSCSC (Robust method extracting simultaneously information from temporal and concentrational domains, i.e. taking all kinetic series from all samples together). This method is more general and theoretically more powerful than PARAFAC as it allows for both "matrix effect" incorportation via having whole matrix of life-time parameters $\tau_{m,k}$, where m stands for species and k for sample. Index i here corresponds to doubles (t_i, k). Expression for modelled luminescence amount is product of (11) and (13). The factor Ω_m (12), unlike for FATS, shouldn't be omitted here (as it depends on life-time $\tau_{m,k}$ which change with k and therefore with i)

$$C_{im}^{(\text{model})} = \\ = C_m^{(speciation)}\left(\beta_m^{spec}; c_{L,k}\right)\cdot\Omega_{im}\exp\left(-\tau_{m,k}/t_i\right) \quad (14)$$

2.2.3 χ^2 minimization procedure

The objective function χ^2_{MEM} can be minimized with constrains when necessary, (e.g. in case of multicomponent analysis of noisy spectra), the constrains can be putted on

i) Individual component spectra (usual constrain should be positivity evaluated in small set of spectral points – this leads to linear inequality conditions on rows of variable matrix R)
ii) C-model parameters (usual constrain should be for life-times τ_m, or speciation constants β_m to be positive or from given interval).
iii) Variance model parameters γ (positivity)
iv) Less usually, the "projected" or "true" luminescence amount profiles C_{im} should be constrained to be positive. C_{im} are evaluated via

$$C = VWR^{-1}. \quad (15)$$

Since (15) is non-linear in R_{mj} original variables, this constrain will lead to series of non-linear inequalities partitioning candidate set/space of parameters (R,β,γ) into several components of continuity.

In the most simplified version, the model of σ_{ij}^2 variance is fixed within minimization and the second, logarithmic term, in objective function (9) can be omitted. This corresponds to an usual χ^2-minimization.

Please note the difference between $C_{im}^{(\text{model})}$ and C_{im} as those elements have in general different values, former corresponding to fitting function value, latter to measurement outcome.

Important note should be made on relationship between the considered factor dimension f and number of individual components b. In an ideal case $f=b$, matrix R is square and (in an ideal subcase) regular. But there is possibility to consider b individual component spectra to be expanded into f subspectra, with $f>b$. In that case R is rectangular and matrix inversions in formulae above has to be taken as pseudoinverses. In the limit $f=s$ and without constrains this lead to simple spectral "deconvolution".

Pre-last note on the three (1), 2), 3)) SVD-based methods should be made on normalization of individual component spectra obtained from R matrix elements (7).

a) FATS and PARAFAC: norm of Z columns ($Z_m(\lambda_i)$, m fixed) corresponds to "luminescence amount" emitted by m-th component and will be denoted $\zeta_{k,m}$ (in case prefactor Ω_m is considered in the model (11), $\zeta_{k,m}$ has dimension s^{-1} (counts per second)). The $\zeta_{k,m}$ is a product of molar concentration[a] $C_{k,m}$ [$\text{mol}\cdot\text{dm}^{-3}$] and „molar luminescence" μ_m [$\text{s}^{-1}\cdot\text{mol}^{-1}\cdot\text{dm}^3$] of a given species,

$$\zeta_{k,m} = \mu_m \cdot C_{k,m}. \quad (16)$$

It is not possible to conclude the two factors on left side of (16) separately.
It might be possible if $s > b$ samples with the same b species would be analyzed and least-square fitting of series of (16) equations (for different k and m) would be done (based on assumption that molar luminescence for the same species m is independent on solution index k[b]), but author has rather negative experience with such a fit, except $s >> b$ [Bc,Mgr].

b) FACSC and FATSCSC: norm of Z columns corresponds to molar luminescence directly. These methods provide direct access to molar concentrations of chemical species in question.

The choice of norm is discussed in 2.2 section of [1]. The raw spectra $Z_m(\lambda)$ (or $Z_m(\nu)$ when considering the wave-number, ν, scale cm^{-1}) are divided by their norms to provide normalized individual component spectra

---

[a] I apologize to dear reader for a confusing notation. $C_{k,m}$ here is different quantity than $C_{i,m}$ in (11) or (15). In (11), we consider time-dependent (hence index i, for FATS parametrizing temporal domain, via $i \to t_i$) concentration of excited state (of species m). But in (16) molar concentration of the respective species m is denoted $C_{k,m}$ (hence index k denoting solution) – i.e. time-independent constant.

[b] And that some additional information is known, here it could be a total uranium concentration. Then,

$$c_k^{(U,\text{tot})} = \sum_m \frac{\zeta_{k,m}}{\xi_m^{(\text{molar})}}, \quad (17)$$

for $k \in \{1, 2, \ldots, s\}$ for $s > b$ is, in principle, ready for Σ (LHS-RHS)^2 fit with variables μ_m to be determined (and subsequent use in (16) to determine $C_{k,m}$ from $\zeta_{k,m}$).

$\zeta_{k,m}$ (and μ_m) contains device-dependent prefactor (independent on m and k) and only their ratios are comparable across literature.

$Z'_m(\nu)$[c]. Their further processing will be discussed in later section 2.2.6

The data preprocessing preceding the SVD and question of factor dimension, *f*, choice will be briefly addressed inside the experimental data analysis section.

2.2.4 PARAFAC

Parallel Factor Analysis (PARAFAC) in this study works with decomposition of order=3 data tensor[d] $Y(\nu,t,k)$ (from all kinetic series of all samples together) according to equation

$$Y(\nu,t,k) = \sum_{m=1}^{N} Z_m(\nu)\cdot D_m(t)\cdot C_m(k) + \varepsilon, \quad (18)$$

where ν is wave-number, t delay and k sample index. The total number of signal component N is free parameter of the method, but can be decided by numerous diagnostics, among them, core consistency, CORCONDIA, available in MATLAB code should be noted. e is deviation which is minimized by the PARAFAC fitting procedure, m index components which are ordered by total luminescence amount connected with them (e.g. D_m could be normalized that $D_m(t_1) = 1$, $Z_m(\nu)$ that euclidean norm is 1 and total luminescence amount of m-th component can be expressed as a norm of $C_m(k)$ profile).

In contrast to SVD-based methods listed in 2.2.2, PARAFAC need neither model of luminescence decay nor speciation model as an input. It is widely known and well utilized method available in several different software implementations and could be therefore used almost as a black-box. This makes it method of the first choice for several preliminary data analyses.

However, aside of the „matrix effect" neglection drawback discussed already in 2.2.2 (point 3)), there is another one – PARAFAC, in its original formulation, needs kinetic series of all samples to be measured with exactly the same temporal point choice (even if some samples exhibit only short-lived luminescence and some long-lived only). This is not case of FATSCSC. For a deeper analysis of complicated systems, PARAFAC results should be taken with caution and rather as second to FATSCSC or FATS results.

2.2.5 Individual component spectra fitting and further analysis

The (normalized) individual component spectra has been fitted to linear combination of seven Gaussian peaks (indexed by index n, which can be interpreted as difference between vibrational quantum numbers of p' → $p, n=p-p'$ in the simplest model either $p'=0$ or $p=0$ and $n \geq 0$ corresponds to cold bands and $n < 0$ to hot bands),

$$Z'_m(\nu) = \sum_{n=-N_h}^{N_c} c^2_{n,m} \exp\left(-\frac{(\nu-\nu_{n,m})^2}{2\sigma^2_{n,m}} \right), \quad (19)$$

where $\nu = 1/\lambda_l$ (connecting $Z'_m(\nu)$ and Z'_{lm} notation) is wave-number, summation limits are $-N_h$ (N_h being number of hot-bands) and N_c (number of cold-bands), $c_{n,m}^2$, $\nu_{n,m}$ and $\sigma_{n,m}$ are n-th peak height, maximum and variance parameter respectively.

Subsequently, peak maxima are correlated with their number n according to formulae

$$\nu_{n,m} = T_{00,m} - \omega_{gs,m}\cdot n, \quad (20)$$

$$\nu_{n,m} = T_{00,m} - \omega_{es,m}\cdot n. \quad (21)$$

Area under n-th peak according to (19)

$$S(n) = \sqrt{2\pi}\,\sigma_{n,m} c^2_{n,m}, \quad (22)$$

could be used to determine ΔR_m through fitting to linear harmonic oscillator Franck-Condon factor ratio as suggested in [1] (page 5).

For single mode, linear harmonic oscillator Franck-Condon factor explicit formula (23) from [FC] has been used. In following formula (23) in this study, simplified version for vibrational quantum number v = 0 has been used (with v' any natural number), i.e.

$$2^n n! \cdot \left|\langle 0;\omega,R|n;\omega',R'\rangle\right|^2 = \left(\sum_{k=0}^{\lfloor n/2 \rfloor} \binom{n}{2k} \left(\frac{4\omega}{\omega+\omega'}\right)^k (2k-1)!!\,H_{n-2k}\left(-\Omega\omega^{-1/2}d\right) \right)^2, \quad (23)$$

with

$$\Omega = \frac{\omega+\omega'}{\omega'\omega}, \quad (24)$$

where $\left|\langle 0;\omega,R|n;\omega',R'\rangle\right|^2$ is the Franck-Condon factor ($\Delta R = |R-R'|$, ω and R are shown to stress that bra and ket vectors from this expression are not dual to each other except for $R=R'$ and $\omega=\omega'$ case). For setting $\alpha = \omega_{gs}$, $\alpha' = \omega_{es}$ (or in reverse order for hot bands) in (23) [FC], the variable $d = C\cdot\Delta R$, where

$$C^2 = \frac{\mu m_u c}{\hbar}, \quad (25)$$

where c is speed of light in vacuum, m_u atomic mass unit (Dalton) and μ reduced mass of vibrational mode in question (here the symmetric stretching mode of uranyl group, i.e., $\mu = m(^{16}O)$ for the most common isotopologue $^{238}U^{16}O_2^{2+}$).

2.2.6 Two or one hot band? Interpretation questions

Since the differences between wgs and wes are even smaller than the ~160 cm^{-1} for $[UO_2(H_2O)_5]^{2+}$ [1] (for $M_nUO_2(CO_3)_3^{2n-4}$ (M=Ca, Mg), the difference $\Delta\omega$ can be less than 50 cm^{-1}), it is hard to determine the crossing point of the two linear branches on peak maximum= *f*(peak number) curve and therefore decide whether studied luminescence spectra exhibit one hot band (and T_{00} ~ 20 000 cm^{-1}) or two hot bands (and T_{00} ~ 20 800 cm-1 differ by one vibrational quantum wgs ~ wes ~ 800 cm^{-1}). The impact to goodness of fit according to Franck-Condon factor formula (23) is greater and the two hot bands model have been found as better consistent with experimental data.

For comparison, the uranyl – sulfate system TRLFS spectra measured under ambient conditions exhibits one

[c] The primes are later dropped and „normalized" is omitted in naming as long as it is not important in particular.

[d] The terms 3-way data or 3-mode data is also widely used in chemometrics. The PARAFAC method can be formulated also for M-way data when $M > 3$.

hot-band only [FOG]. Why would cryogenic conditions lead to greater number of hot bands (and much greater portion of luminescence emitted in the hot band peaks)? Possible answer might be that deexcitation in solid phase sample doesn't enter „thexi" stage as in liquid case and vibrationaly excited substrates of electronic excited state are therefore stabilized. To assure both two hot band interpretation and theoretical explanation of its origin, series of TRLFS measurements on uranyl – sulfate system in both aqueous solutions and ice under several different temperatures should be done.

2.2.7 Identification of given chemical species–individual component assignment problem

Neither FATS nor PARAFAC could provide us with definite answer on chemical composition of samples in question. After the data analysis, we are left with a list with rows τ_m, ζ_m, $(T_{00,m}, \omega_{gs,m}, \ldots)$ for $m \in \{1,2, \ldots, f\}$ only. The interpretation to which chemical species $m = 1$, $m = 2$, …, $m = f$ components correspond is yet to be done. There are several possibilities for the above-mentioned assignment:

a) Literature search for experimental spectra
b) Experimental speciation study (cryo-TRLFS[e] measurement on series of artificial samples (solutions made from pure chemicals as UO_3, Na_2CO_3, Na_2SO_4, $CaSO_4$, …))
c) Quantum-chemical modelling (self-made or literature search).
d) By comparison of PARAFAC obtained luminescence-speciation and geochemical model (provided, e.g. by PhreeqC modelling based on existing thermodynamic properties database).

3 Experimental

3.1 Sample characterization

… sample origin
… how S3 -> S5 -> S4, acidification
… total elemental analysis

Following Table 1 presents total elemental concentrations in mg/l, pH and EH in mV, sample number is written in the first row. S(VI) stands for sulfate SO_4^{2-}, C(IV) for hydrogencarbonates (bicarbonates) HCO_3^- and carbonates (CO_3^{2-}, except for S9 almost all C(IV) is in the form of HCO_3^-), N(V) for NO_3^-.

Table 1: Total elemental conc. in mg/l

E \ S	1	2	3	4	5	9
Na	121	122	1350	1550	1580	2520
K	12.8	12.8	27.4			
Mg	104	100	301	333	340	61.8
Ca	158	191	292	296	302	86.5
Fe	5.09	0.18	0.28			0.08
Mn	2.17	1.4	0.482			
U	1.68	0.02	3.1	3.38	3.53	10.8
Cl	54.3	452	496	908	815	1170
S(VI)	576	546	3620	4030	4100	3150
C(IV)	574	14	657	113	300	2244
N(V)	<0.5	<0.5	3.6			
pH	7.1	7.4	7.3	5.5	6	9.3
E_H	130	450	310	310	310	400

[e] Time-Resolved Laser-Induced Fluorescence Spectroscopy [TRLFS]. The phenomenon in given samples investigated should be, however rather called "luminescence" since the electronic excited state producing measured light is mixture of triplet and singlet state.

3.2 Measurement

The measurements have been done by HZDR collective authoring previous report on the same topic [] and detailed experimentals can be found there []. Important information

… sample cooling and temperature
… laser type
… background subtraction
… temporal axis sampling, wave-length sampling

4 Data analysis of Experimental results

4.1 FATS computationals

The SVD decomposition as described in 2.2.1-2.2.3 has been applied with weighting-preprocessing such that original measurement data matrix elements Y_{li} has been transformed onto Y'_{li} normalized data matrix via

$$Y'_{li} = \frac{Y_{li}}{P_l Q_i}, \tag{26}$$

where $P_l Q_j$ represent separable form of variance of Y_{li}, so Y'_{li} are now closer to case of independent and identicaly distributed random variables. Since greater variance lies along temporal domain, Pl can be set as $P_l = 1$ and Q_i has, in software MyExpFit, general form

$$Q_i = \underset{l}{mean}|Y_{li}|^{\phi}, \tag{27}$$

where $\phi \in (0;1)$. For purely poisson noise, $\phi = 0.5$, this choice has been applied here. After SVD procedure (4), (5), $Y' = U'WV'^T$, the subspectra and loadings should be „denormalized" back according to formulae below,

$$U_{lj} = U'_{lj} P_l, \tag{28}$$

$$V_{ij} = V'_{ij} Q_i. \tag{29}$$

The factor dimension f can be, in general case, determined according to three different diagnostics:

i) SCREE-plot diagnostics focus on crossing of linear branches in log $W_j = f(j)$ plot (Fig. 1). On example for sample S3, f=3 or f=4 since the last branch with smallest slope corresponds to noise, but from SCREE-plot alone it is hard to recognize whether still to include $j = 4$ component or not.
ii) Loadings-based V-diagnostics focus on number of first, signal-like, loadings. For S3 sample, Fig. 2 and Fig. 3 show three signal-like loadings. Since fourth, loadings are much more noise-like (Fig. 4).
iii) Subspectra-based U-diagnostics works as previous, except for subspectra. Fig. 5 presents first three signal-like subspectra, Fig. 6 subspectrum U4' already noise-like.

While ii) and iii) diagnostics for chosen example (Sample S3 TRLFS kinetic series analysis) suggest to accept $f = 3$ components for further analysis, it is better just to conclude that it is possible to statistically distinguish $N = 3$ independent luminescence active

DOI: 10.1051/epjconf/201715401029

species (according to the geochemical modelling it should be $Ca_2UO_2(CO_3)_3^0$, $CaUO_2(CO_3)_3^{2-}$ and most probably $MgUO_2(CO_3)_3^{2-}$), but set $f = N + 1 = 4$.

Because the software based background correction/subtraction is rather approximate, irrespectively to method free parameter choice, FATS for $f > 2$ provide one individual component with several orders of magnitude larger life-time (and smaller luminescence amount) and bandless continuum-like luminescence spectrum. This component corresponds to background artifact and is ignored in further analysis. Therefore, for N chemical component model, $f = N + 1$ has to be chosen. By preliminary analysis of randomness of residuals, case $N = 1$ has been found as insufficient for any sample investigated below, $N = 2$ as a slight under-fit and $N = 4$ as a slight over-fit.

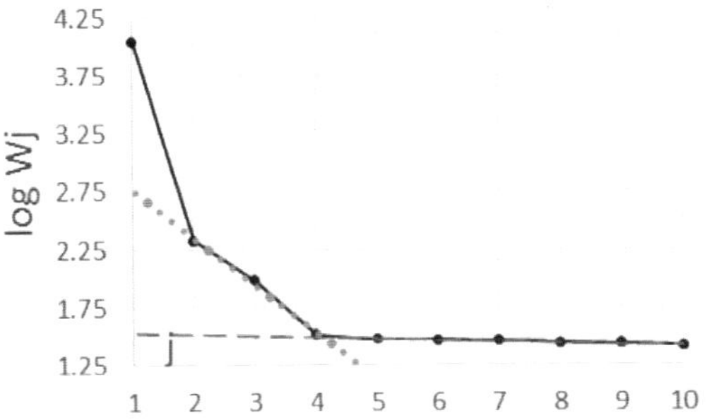

Fig. 1. SCREE plot for NmSVD of connected TRLFS kinetic series for sample S3. $W_j = W_{jj}$ is the j-th singular value, j on horizontal axis. The singular values are plotted as black circles connected by thick lines. Dashed and dotted lines correspond to linear approximation of j = 5 to 10 and j = 2 to 4 regions, respectively.

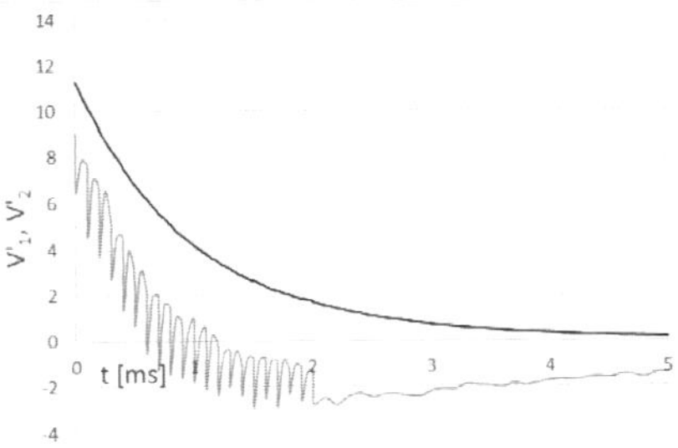

Fig. 2. NmSVD Loadings (sample S3) of the first two components (j = 1 (black) and j = 2 (gray)).

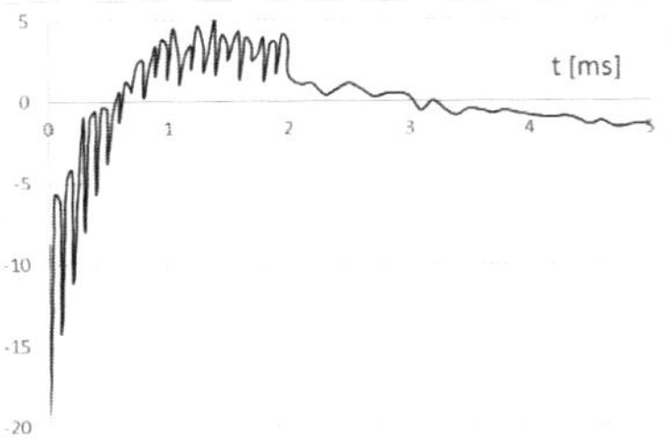

Fig. 3. NmSVD Loadings (sample S3) of third component (j = 3, still considered as signal).

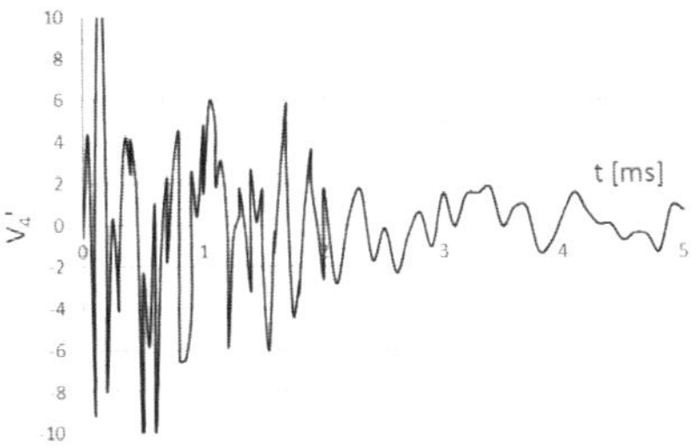

Fig. 4. NmSVD Loading (Sample S3) of fourth component – very noisy, represent background and noise. Last loading taken into further analysis.

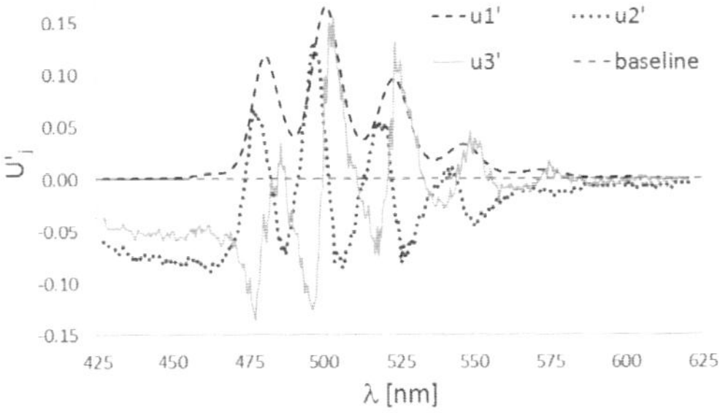

Fig. 5. NmSVD subspectra corresponding to the first three components ($j \in \{1,2,3\}$), clearly corresponding to subspace of signal (in R^N, where N = 419 is number of spectral points selected).

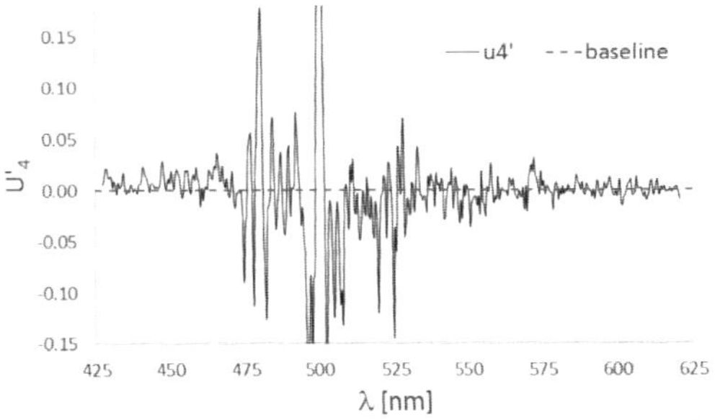

Fig. 6. First, NmSVD subspectrum U'_4 corresponding to noise has been added to further analysis for the need to represent background subtraction artifact, but subspectra for j > 4 are omitted from further analysis.

4.2 Sample S1 (m-F510)

NmSVD for this sample suggest N = 3 (or 2) and therefore f = 4 (or 3) (Fig. 7). Denormalized f loadings from NmSVD has been fitted to linear combination of f exponential decays, resulting luminescence amounts ζ_m [10^3 ms^{-1}] and life-times τ_m [ms] are presented in Tab. 2.

FATS f=3 and f=4 analysis has been done. Resulting individual component spectra (Figs. 8, 9) in the former case contained "structure" inside the background component (dotted in Fig. 8) which favourized latter f parameter choice. Please note the peak maxima (e.g. for the highest peak) of the component differs by a tiny portion of 0.86 nm (less than twice of wave-length sampling period !) for f=3 and ~1 nm (~50 cm^{-1} in this region) for the two charged complex species for $f = 4$ as well.

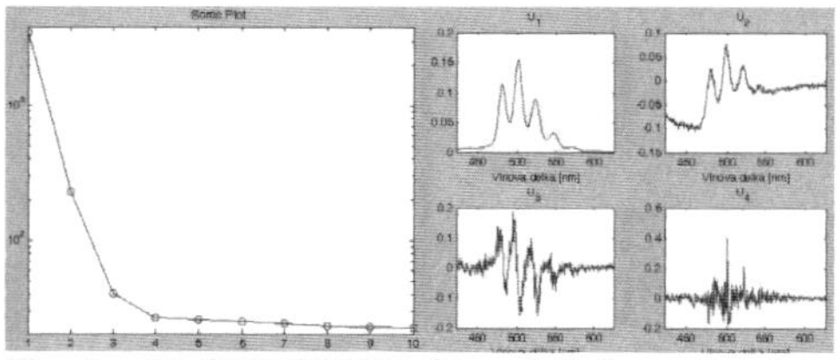

Fig. 7: NmSVD SCREE-plot and U-diagnostics done by MyExpFit V4 program.

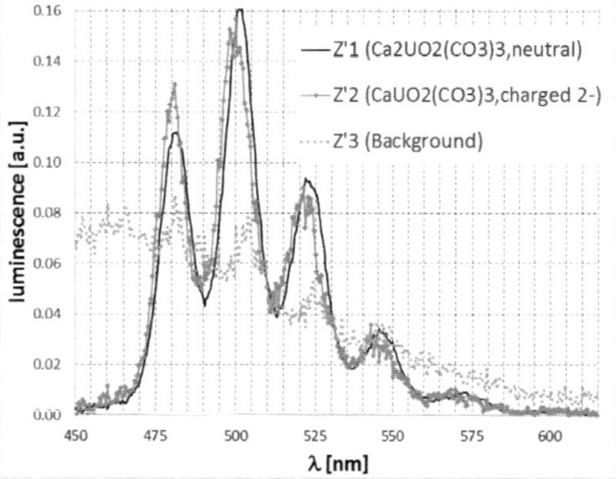

Fig. 8. FATS, $f = 3$ analysis of sample S1 (m-F510) results for individual component spectra (nm scale).

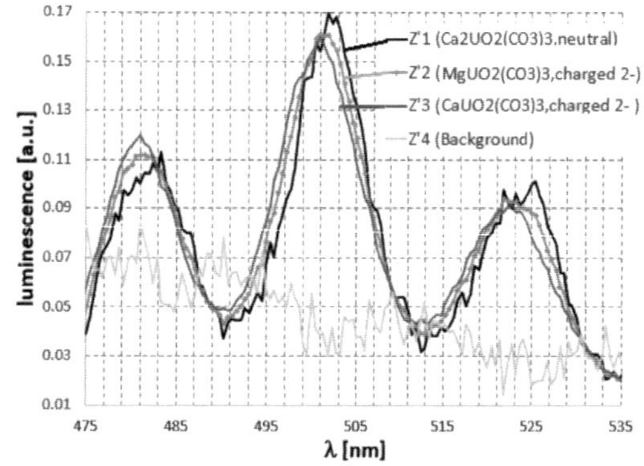

Fig. 9. FATS, $f = 4$ analysis of sample S1 (m-F510) results for individual component spectra (nm scale, for better visibility of peak position differences, only three largest peak detail is shown).

Spectroscopic parameters – T_{00} [10^3 cm^{-1}] (excitation energy) and ω_{gs} [cm^{-1}] (ground state vibrational frequency) have been calculated from gaussian fits of FATS $f = 4$ individual component spectra (Tab. 2, Fig. 10, 11 as an example for $m = 1$ component).

For this sample example of ΔR estimation from experimental data is shown in Fig. 11 and Fig. 12, where one and two hot band models, respectively, have been used. In their comparison, two hot band model seems to be more realistic with $\Delta R = (7.3 \pm 1.0)$ pm

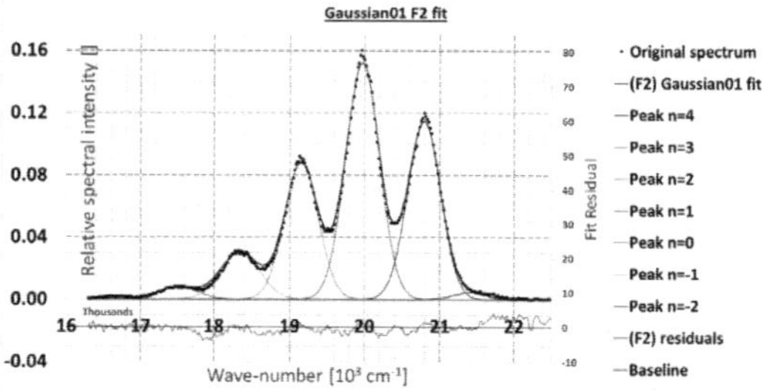

Fig. 10. $m = 1$, $\tau_1 = 0.87$ ms component (assigned to $MgUO_2(CO_3)_3^{2-}$) of $f = 4$ FATS decomposed according to (19) (cm^{-1} scale).

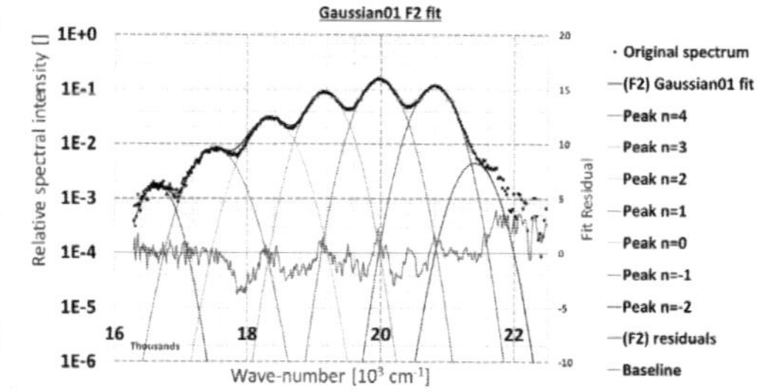

Fig. 11. $m = 1$ component ($f = 4$), intesity plotted in logarithmic scale to show approximate gaussian character for red-most and blue-most tiny peaks too.

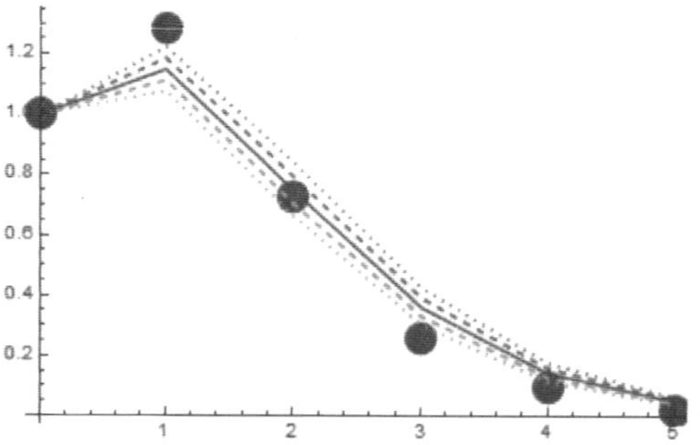

Fig. 12. FATS $f = 4$, $m = 0$ ($\tau_0 = 0.36$ ms, assigned to $CaUO_2(CO_3)_3^{2-}$) Franck-Condon fit (23) within „one hot-band" model. Fitted $\Delta R = (10.4 \pm 1.0)$ pm. On vertical axis, $n' \to 0$ peak area (22) relative to $0' \to 0$ peak is plotted. Horizontal axis represents vibrational number n'.

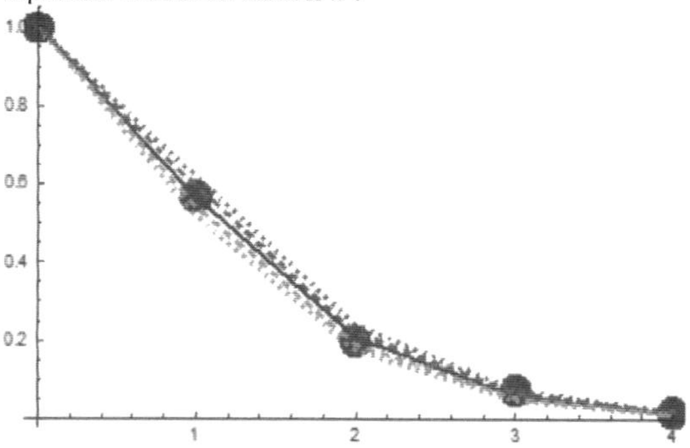

Fig. 13. FATS f = 4, m = 0 as in previous Fig. 11, but within „two hot bands" model. $\Delta R = (7.3 \pm 1.0)$ pm.

Table 2: FATS $f = 4$ results for sample S1.

m	τ_m	ζ_m	$T_{00,m}$	ω_{gs}
0	0.36	8.1	20.01 ± 0.01	794
1	0.87	30.1	19.97 ± 0.02	816

2	1.52	4.51	19.92 ± 0.03	807
3	9.52	0.09	n.d.	n.d.

4.3 Sample S2 (m-555)

Same procedure as for S1 has been repeated, results are presented in Table 3 below (since now, background component is not listed). The uncertainities are d(wgs) ~10cm-1, d(tm) ~ 0.1 ms and d(zm) ~ 0.2 ms-1.

Table 3: FATS analysis results, sample S2 (m-555)

f	m	τ_m	ζ_m	$T_{00,m}$	ω_{gs}
3	0	0.45	2.57	19.98 ± 0.01	804
	1	1.15	2.53	19.94 ± 0.02	833
4	0	0.21	1.83	20.07 ± 0.04	798[f]
	1	0.76	2.38	19.97 ± 0.02	810
	2	1.26	1.55	19.92 ± 0.03	830

4.4 Sample S3 (E-394)

NmSVD and FATS as done for previous samples revealed following parameters (Tab. 4). Due to the higher uranium content, s/n ratio has been higher and smoother individual component spectra resulted (Fig. 14). An example of fitting procedure for T00, and w_{gs}, determination is presented in Fig. 15, the uncertainities are lower than in S1 by roughly factor of 2.

Table 4: FATS analysis results, sample S3 (E-394)

f	m	τ_m	ζ_m	$T_{00,m}$	ω_{gs}
3	0	0.81	93.2	19.98 ± 0.01	817
	1	1.39	36.9	19.93 ± 0.01	820
4	0	0.75	39.6	19.99 ± 0.01	810
	1	0.88	59.5	19.96 ± 0.01	817
	2	1.44	31.0	19.93 ± 0.01	819

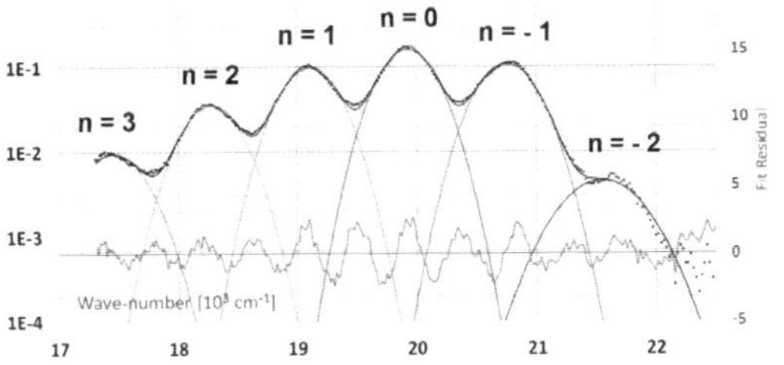

Fig. 14: FATS (f=4), m=2 individual component spectrum (later assigned to $Ca_2UO_2(CO_3)_3^0$) fitted to linear combination of gaussian profiles. Peak maxima have been fitted as a function of peak number n in follow Fig. 15.

4.5 Sample S4 (E-394a)

FATS analysis for f = 4 of the sample created by acidification of S3 by HCl (to pH = 5.5) possibly revealed in previous samples not present component with both high T_{00} and ω_{gs} (m=1, τ_1=1.15 ms), which might be either UO_2CO_3 or UO_2SO_4, but its small luminescence amount (ζ_1~0.07·10[3] ms^{-1}, Table 5) is in contrast with high relative content predicted by geochemical modelling for this acidified sample. Similarly, the component associated with τ=0.84 ms (f=3) or τ=0.88 (f=4) could be assigned to UO_2CO_3 (ω_{gs}=815 cm^{-1}, T_{00}=20080 cm^{-1} fitted from deep cryogenic measurements done in [WAN/ZAC2004], τ=0.47 ms) or UO_2SO_4 due to the previously reported luminescence life-time of UO_2SO_4 under cryogenic conditions of 0.4 ms.

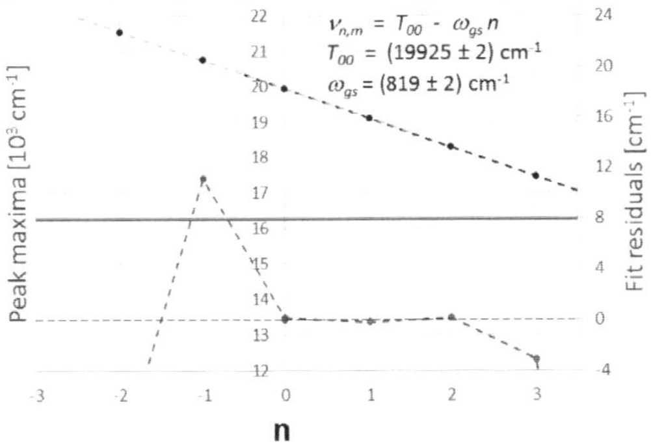

Fig. 15: Upper part: Peak maxima as peak number function (vertical principial axis in 10^3 cm^{-1} divides the two linear branches with almost identical slopes, the uncertainity of fit underestimates total experimental uncertainities), lower part: residuals of the fit (right axis in cm^{-1})

Table 5: FATS analysis results, sample S4 (E-394a)

f	m	τ_m	ζ_m	$T_{00,m}$	ω_{gs}
3	0	0.83	36.9	19.99 ± 0.01	816
	1	1.34	24.3	19.92 ± 0.01	819
4	0	0.83	36.6	19.99 ± 0.01	815
	1	1.15	0.07	20.02 ± 0.02	827
	2	1.34	24.4	19.92 ± 0.01	819

4.6 Sample S5 (E-394b)

For the uranyl-sulfate system, please see [7] and [96,36], where values of stability constants have been corrected.

Uranyl-selenate

Table 6: FATS analysis results, sample S5 (E-394b)

f	m	τ_m	ζ_m	$T_{00,m}$	ω_{gs}
3	0	0.84	97.4	19.96 ± 0.02	816
	1	1.44	34.4	19.92 ± 0.02	820
4	0	0.59	6.81	20.06 ± 0.02	800?
	1	0.88	98.5	19.95 ± 0.02	816
	2	1.51	26.9	19.91 ± 0.02	830

Uncertainities at least 20 cm-1.

4.7 Sample S9 (M-207A)

For the uranyl-sulfate system, please see [7] and [96,36], where values of stability constants have been corrected.

Uranyl-selenate

Table 7: FATS analysis results, sample S9 (M-207A)

f	m	τ_m	ζ_m	$T_{00,m}$	ω_{gs}
4	0	0.48	87.7	19.99 ± 0.01	799
	1	0.91	225.3	19.97 ± 0.01	806
	2	1.50	134.4	19.93 ± 0.01	807

[f] However after not one but two redmost peaks are excluded from the fit, ω_{gs} = 833 cm^{-1} value results.

4.8 PARAFAC analysis of all samples together

PARAFAC decomposition according to (18) led to luminescence amount profiles (Fig. 18), by maximizing objective function

$$f(\pi) \equiv \sum_{m=1}^{N}\left(\sum_{k=1}^{s}\Gamma_{k,m}M_{k,\pi(m)}\right)W_m, \qquad (30)$$

where $\Gamma_{k,m}=C_k(m)/\Sigma_n C_k(n)$ is relative luminescence amount of m-th component (18) in k-th sample, $M_{k,\pi(m)}$ is $\pi(m)$-th species concentration in k-th sample according to geochemical modelling calculation (Fig. 19, done in PhreeqC [], []), W_m is weight, N = 7 (the background component is excluded) and maximization is done with respect to permutation π. Resulting optimal assignment is presented in Table 8 and parameters τ [ms], T_{00} [cm^{-1}] and ω_{gs} [cm^{-1}] (from monoexponential fits of $D_m(t)$ and gaussian fits of $Z_m(\nu)$ (18)) should be compared to found literature experimental data (Table 9) and quantum chemical calculations (section 6).

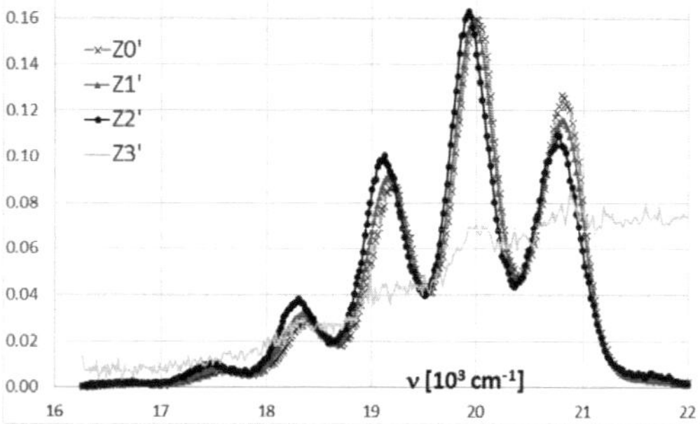

Fig. 16.

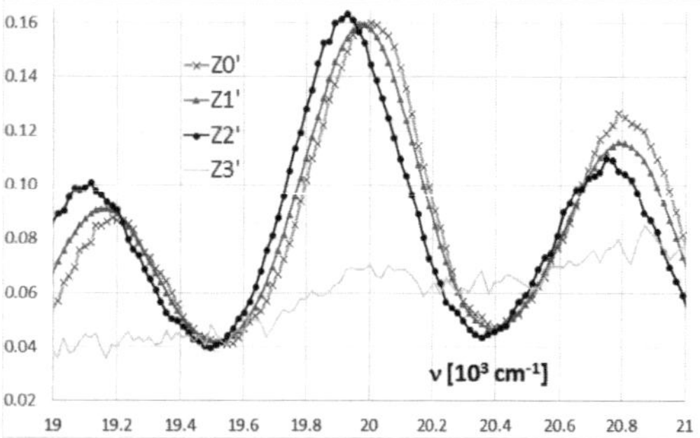

Fig. 17.

Table 8: Optimal component assignement

m	Species	[illegible]	T_{00}	ω_{gs}
1	$CaUO_2(CO_3)_3^{2-}$	0.551	20051	806.0
2	$Ca_2UO_2(CO_3)_3$	1.322	19916	823.4
3	$UO_2(CO_3)_3^{4-}$	1.001	19858	812.3
4	$MgUO_2(CO_3)_3^{2-}$	0.902	19973	797.1
5	UO_2CO_3	1.091	20089	816.9
6	UO_2SO_4	0.658	19970	831.4
7	$UO_2(CO_3)_2^{2-}$	1.044	19914	798.4

Table 9: Comparison with literature

m	$T_{00,REF}$	$\omega_{gs,REF}$	τ_{REF}	REF
1			0.615	
2	19943	798 ;820	1.009	
3	19995		0.887	
5			0.905	
	20081	815	0.465	
6			0.400	
7			0.144	
	20145	810	0.963	

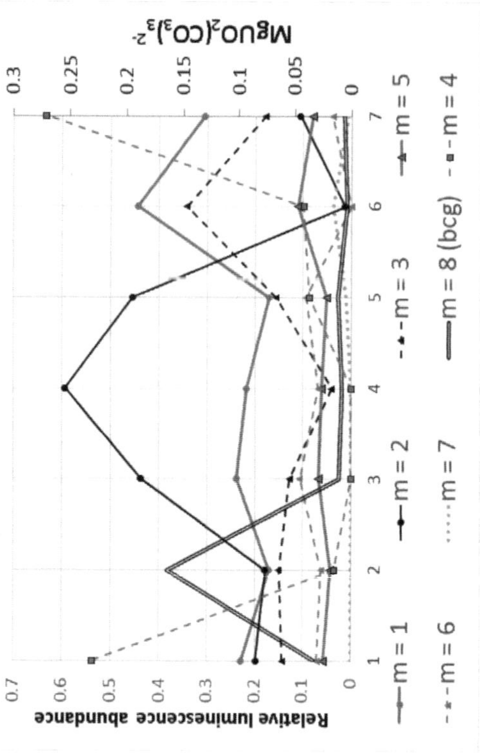

Fig. 18.

5 Computationals for quantum chemical calculations

Similar to [1] calculations were done in Turbomole V6.5 program. Scalar relativistic effects were approximately included via 60 electron Effective Core Potential (ECP) for uranium atom, all other electrons were, at this level of approximation treated as non-relativistic. Hydration has been described by explicit inclusion of small number (denoted n in chemical formulae like $[UO_2(\kappa^2\text{-}CO_3)(H_2O)_3].nH_2O$) of water molecules into investigated system. The atomic basis sets used were def-SVP (denoted „S" in all tables in the section 6 below, [30,100]) and def-TZVPP (denoted „T", [29,30], for solvated models, for solvent water molecules only def-SVP atomic basis set was used to reduce the computational demands).

For the ground state properties DFT/B3-LYP [28,82-85] has been used. For DFT calculations the D3 dispersion correction was applied [93]. All figures have been exported from Molden [104] screenshots.

 DOI: 10.1051/epjconf/201715401029

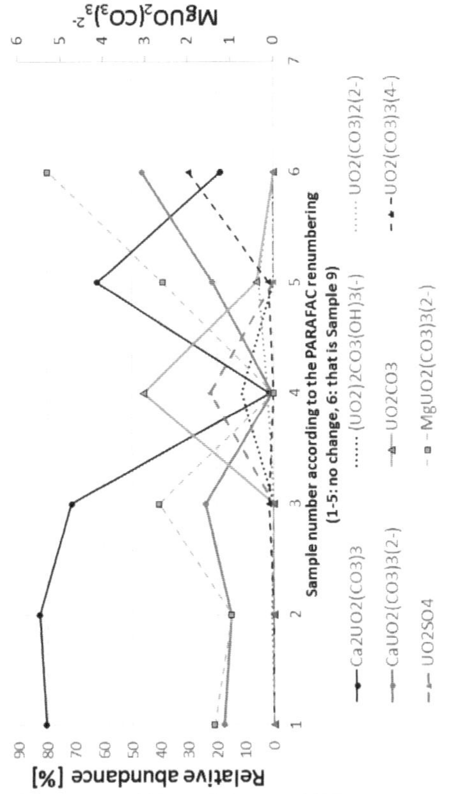

Fig. 19.

6 Quantum chemical calculation results

6.1 Uranyl Monocarbonate, UO_2CO_3

Similarly as for the uranyl monosulfate, UO_2SO_4, four different configurations should be investigated, differing by dentacity of CO_3^{2-} ligand (κ^2- for bidentate, coordinated by two donor oxygen atoms, κ^1- for monodentate, coordinated by one donor oxygen atom) and coordination number of central uranyl group UO_2^{2+} (i.e. number of water molecules coordinated to U along carbonate ligand), labeling from [1] will be adopted leading to κ^2co5 (for $[UO_2(\kappa^2\text{-}CO_3)(H_2O)_3]$), κ^2co6 ($[UO_2(\kappa^2\text{-}CO_3)(H_2O)_4]$), κ^1co5 (for $[UO_2(\kappa^1\text{-}CO_3)(H_2O)_4]$) and κ^1co6 (for $[UO_2(\kappa^1\text{-}CO_3)(H_2O)_5]$). In vacuum, only κ^2co5 has been found stable, explicit inclusion of water molecules allowed stabilization sligthly less stable κ^2co6 and when at least 18 water molecules have been added around studied complex, also κ^1co5. Ground state symmetric stretching mode frequency of uranyl group ω_{gs} and vertical (T_e, not T_{00}) excitation energies are reported in Table 10 below, where b. stands for atomic basis set, n for number of explicit water molecules (not coordinated to uranyl, but as solvent), *E* for relative energy (in mH for different b., in cm^{-1} if bold - for different local PES minima (b.,form and n identical), relative to closest most „0" upside in the table), *R* for U-O_{yl} bond distance in pm.

Table 10: UO_2CO_3 prop. quant. chemical estimate.

ref	b.	form	*n*	*E*	T_e	ω_{gs}	*R*
1	S	k2co5	0	808	18808	873	179
2	T			38	18611	865	178
3	Q			0	18631	866	178
4	S	k2co5	12	**0**	20703	850	179
5	S		12	**3206**	20655	857	179
6	S		12	**3505**	20553	820	180
7	Tc		12	-874	20448	850	179
8	Tc		12		20250	813	180
9	*T*		*12*	*-1927*	*20394*	*856*	*178*
10	S		16	0	20695	865	178
11	Tc		16	-774	20453	868	178
12	S		25	0	21139	856	179
13	*Tc*		*25*	*-773*	*20863*	*852*	*178*
14	S	k2co6	12	**0**	20569	887	177
15	S		12	**1264**	20588	901	176
16	S		12	**1657**	20757	881	177
17	Tc		12	-872	20328	886	177
18	Tc		12	-862	20356	896	176
19	S		20			861	178

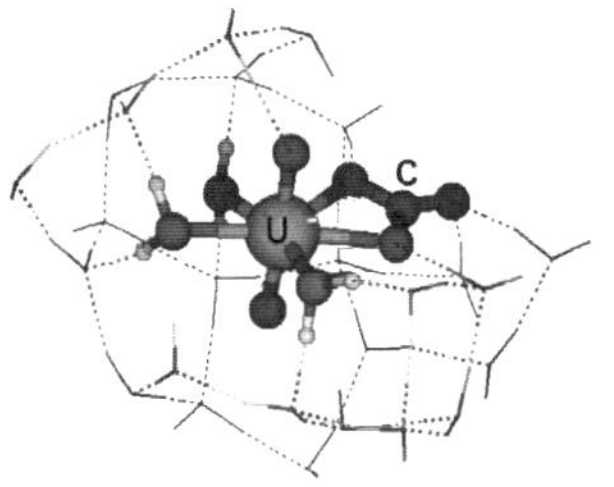

Fig. 20. At def-TZVPP/B3LYP/DFT-D3 level optimized geometry of $[UO_2(\kappa^2\text{-}CO_3)(H_2O)_3]$. 25 H_2O (solvent water molecules are represented by lines and def-SVP basis set us used for their description). Please note the planarity of CO_3^{2-} ligand situated in equatorial plane (approximately perpendicular to uranyl axis)

6.2 Uranyl bis(carboate), $UO_2(CO_3)_2^{2-}$

Among all six possibilities only κ^{22}co5 isomer (supposed to be most stable) has been investigated. Results are presented in table below.

Table 11: Ground electronic state properties of $[UO_2(\kappa^2\text{-}CO_3)_2(H_2O)]^{2-}$ (U-O_{yl} bond length *R* in pm, vibrational frequency of UO_2^{2+} symmetric mode ω_{gs} [cm^{-1}]) and vertical excitation energies T_e [cm^{-1}].

ref	b.	*n*	*E*	T_e	ω_{gs}	*R*
1	S	0	0	20551	814	181
2	T	0	-905	20349	808	181
3	Q	0	-952	20355	810	180
4	S	12	0	20665	830	179
5	Tc	12	-893	20507	827	179

Aside of the vibrational frequencies in Tab. 11, there is also another mode with smaller contribution to uranyl symmetric streching with ω of 723, 710 and 710 cm^{-1} for ref. 1, 2 and 3 respectively.

For this molecule, electronic excited state PES local minima solution has been done and adiabatic energies T_a and T_{00} are presented together with deexcitation energy T_{de} [all in cm^{-1}] in the following table (ω_{es} is the symmetric stretching mode vibrational frequency of uranyl group in the excited electronic state in cm^{-1} and ΔR excitational elongation in pm).

Table 12: Excited electronic state properties of $[UO_2(\kappa^2\text{-}CO_3)_2(H_2O)]^{2-}$.

ref	b.	*n*	T_a	T_{00}	T_{de}	ω_{es}	ΔR
1	S	0	20075	19724	19532	677	3.9
2	T	0	19843	19542	19294	669	4.1
3	Q	0	19854		19312		4.1
4	S	12	20089		19442		

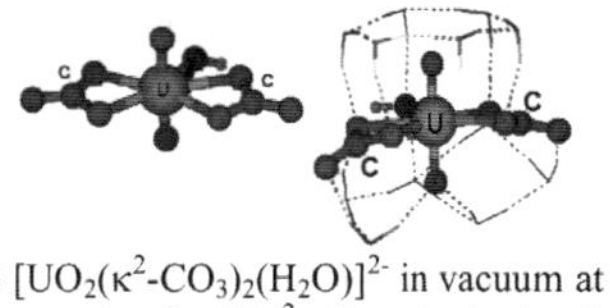

Fig. 21. Left: $[UO_2(\kappa^2\text{-}CO_3)_2(H_2O)]^{2-}$ in vacuum at def-QZVPP atomic basis set level, CO_3^{2-} ligands both planar and in equatorial plane, intramolecular H-bond, right: with 12 water solvent molecules around, carbonate ligands slightly out of equatorial plane.

6.3 Uranyl tris(carbonate), $UO_2(CO_3)_3^{4-}$

For uranyl

Table 13: Ground electronic state properties of $[UO_2(CO_3)_3]^{4-}$ (U-O$_{yl}$ bond length *R* in pm, vibrational frequency of UO_2^{2+} symmetric mode ω_{gs} [cm^{-1}]) and vertical excitation energies T_e [cm^{-1}].

ref	b.	*n*	*E*	T_e	$\omega_{gs,1}$	$\omega_{gs,2}$	*R*
1	S	0	0	22104	780	724	183
2	T	0	-1165	21765	778	714	182
3	Q	0	-1234	22453	780	715	182
4	S	12	0	21434	**829**	720	179
5	Tc	12	-1122	21217	**838**	712	179
7	S	24	0	21290	**867**		178
8	Tc	24	-1106		**863**		178

Table 14: Excited electronic state properties of $[UO_2(CO_3)_3]^{4-}$

ref	b.	*n*	T_a	T_{de}	ΔR
1	S	12	20862	20258	4.2
2	Tc	12	20670	20092	3.9

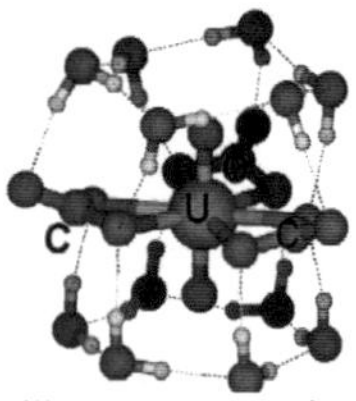

Fig. 22. „Sandwich-like" structure of $[UO_2(CO_3)_3]^{4-}\cdot 12H_2O$ (def-TZVPPc level).

6.4 Ternary complex $MgUO_2(CO_3)_3^{2-}$

Optimized structure without molecules of solvent in presented in Fig. 24. Two CO_3^{2-} ligands are distorted out of equatorial plane and their coordination mode towards uranium is monodentate. Explicit inclusion of solvent molecules allows carbonate ligands to stay in the coordination plane, Fig. 25 shows two possibilities differing by 486 cm-1 (the left one has been found as lower by energy and is presented first in Tab. 15).

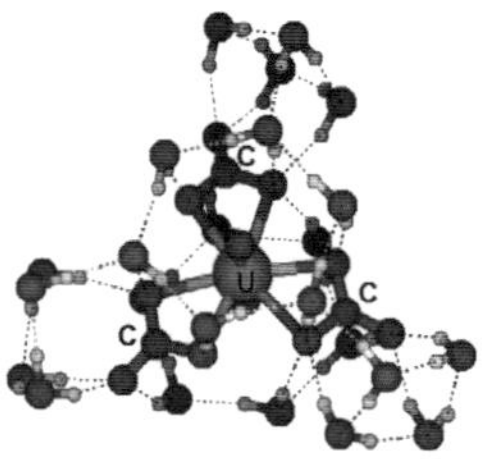

Fig. 23. Struct. corresponding to ref. 8 in Tab. 13.

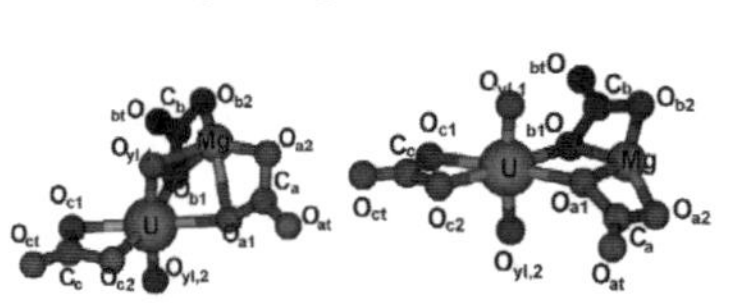

Fig. 24. Optimized structures (def-TZVPP basis) for $MgUO_2(CO_3)_3^{2-}$ in vacuum.

Inclusion of 22 water molecules to model led Mg atom to stabilize closer to equatorial, ligation, plane (Fig. 26-28). Bond lengths are given in Tab. 24 (for description, please see Fig. 24 and Fig. 25)

Table 15: Ground electronic state properties of $MgUO_2(CO_3)_3^{2-}$ (U-O$_{yl}$ bond length *R* in pm, vibrational frequency of modes containing significant part of UO_2^{2+} symmetric stretching mode $\omega_{gs,1}$ and $\omega_{gs,2}$ [cm^{-1}]) and vertical excitation energies T_e [cm^{-1}].

ref	b.	*n*	*E*	T_e	$\omega_{gs,1}$	$\omega_{gs,2}$	*R*
1	S	0	0	19345	778	717	183
2	S	0	757	19845	841	729	180
4	T	0	-1194	19210	769	707	183
5	T	0	-1193	19697	840		179
7	S	12	0	20837	*850*	*700*	183
8	S	12	638	20647	825		180
9	S	12	710	20737	802		181
10	Tc	12	0	20470	*850*	*765*	184
11	Tc	12	486		823		180
12	Tc	12	1326		801		180
13	S	22	0	20854	845		180
14	S	22	396	21210	824		181
15	Tc	22	0	21315	820		180
16	Tc	22	625	21206	826		178

Table 16. Lengths of important bonds in $MgUO_2(CO_3)_3^{2-}\cdot nH_2O$ model structures.

ref	MgO$_{yl,1}$	MgO$_{a1}$	MgO$_{a2}$	UO$_{a1}$	UO$_{a2}$	UO$_{c1}$
4	212	224	193	231	231	228
5		203	194	241	241	229
10	212	202	**MgO$_{b1}$**	260	245	234
11		231	**198**	231		236
15		205	**200**	249	249	241
16		200		239	252	240

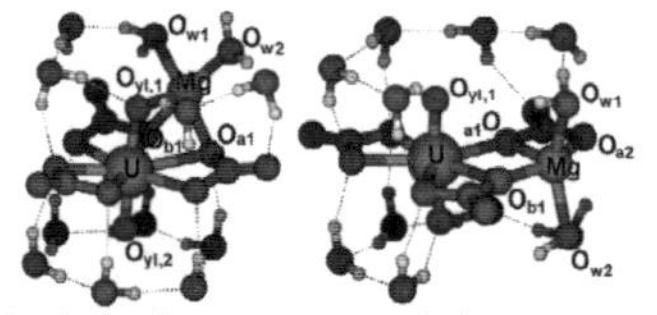

Fig. 25. Optimized structures (def-TZVPPc basis) for $MgUO_2(CO_3)_3^{2-}\cdot 12H_2O$.

Fig. 26. Optimized structure (def-TZVPPc) for $MgUO_2(CO_3)_3^{2-}\cdot 22H_2O$. „Side view". All carbonate ligands are bidentate coordinated and Mg atom is only slightly above coordination plane, its coordination number is 5.

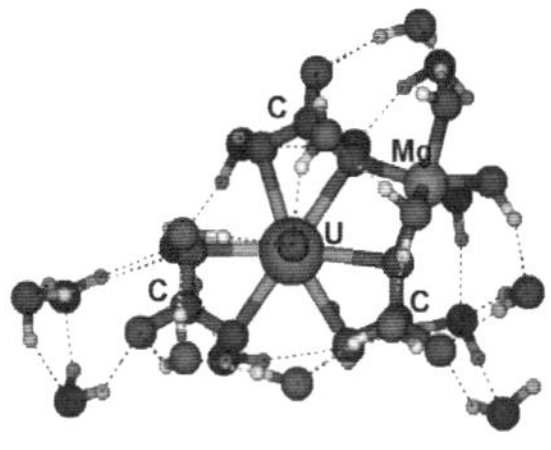

Fig. 27. Optimized structure (def-TZVPPc) for $MgUO_2(CO_3)_3^{2-}\cdot 22H_2O$. „Top view"

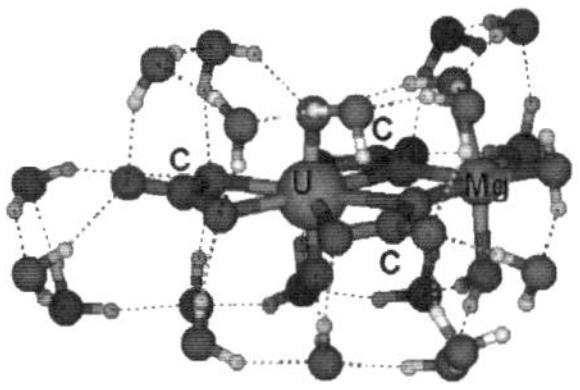

Fig. 28. PES local minimum structure (def-TZVPPc) for $MgUO_2(CO_3)_3^{2-}\cdot 22H_2O$. „Side view". All carbonate ligands are bidentate coordinated and Mg atom position is in coordination plane, with coordination number 6. This structure is 625 cm^{-1} higher than the one from Fig. 27.

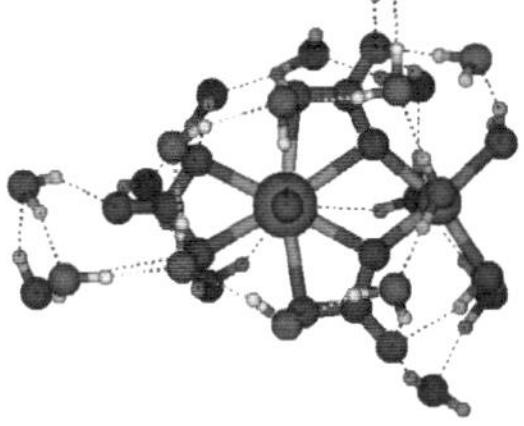

Fig. 29. „Down view" on structure from Fig. 28.

6.5 Ternary complex $CaUO_2(CO_3)_3^{2-}$

Detailed parametric study on number of explicit water included will be given in Supplementary Information. For highest number of explicit water molecules Ca atom preferred position slightly above coordination plane (Fig. 30-32) and tendency to coordination number 7 (Fig. 32).

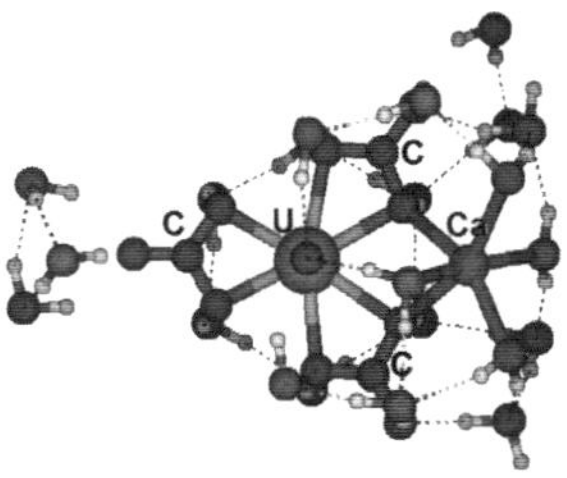

Fig. 30. Top view on optimized $CaUO_2(CO_3)_3^{2-}\cdot 22H_2O$ structure (within def-SVP atomic basis set), associated with ω_{gs} = 797 cm^{-1}, R = 181 pm and T_e = 21256 cm^{-1}.

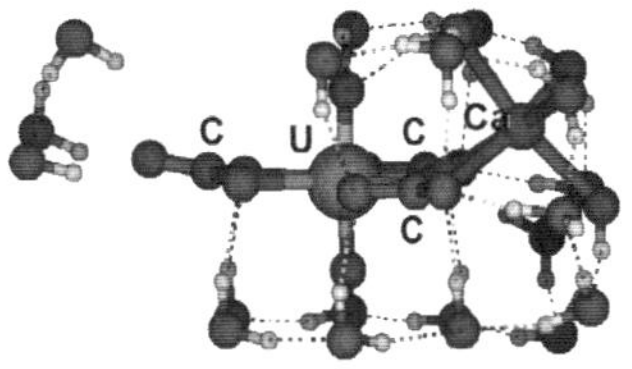

Fig. 31. Side view on optimized $CaUO_2(CO_3)_3^{2-}\cdot 22H_2O$ structure (within def-SVP atomic basis set, see Fig. 30)

Spectroscopic properties are listed in Table 17 below.

Table 17: Calculated properties of $CaUO_2(CO_3)_3^{2-}.22H_2O$ models.

	b.	E	T_e	$\omega_{gs,1}$	$\omega_{gs,2}$	R
1	S		21256	797		181
2	Tc	0		811	794	181
3	Tc	1837cm^{-1}	20475	824		180
4	Tc	2559cm^{-1}		810		181

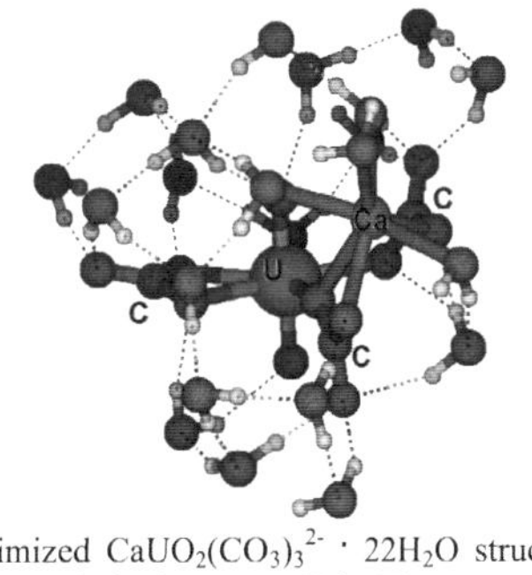

Fig. 32. optimized $CaUO_2(CO_3)_3^{2-}\cdot 22H_2O$ structure (within def-TZVPPc atomic basis set, see Tab. 17)

6.6 Ternary complex $Ca_2UO_2(CO_3)_3^0$

Similar features as in charged complex $CaUO_2(CO_3)_3^{2-}$ has been found for neutral bicalcium analogue $Ca_2UO_2(CO_3)_3^0$ as well (Figs. 33, 34, Tab. 18)

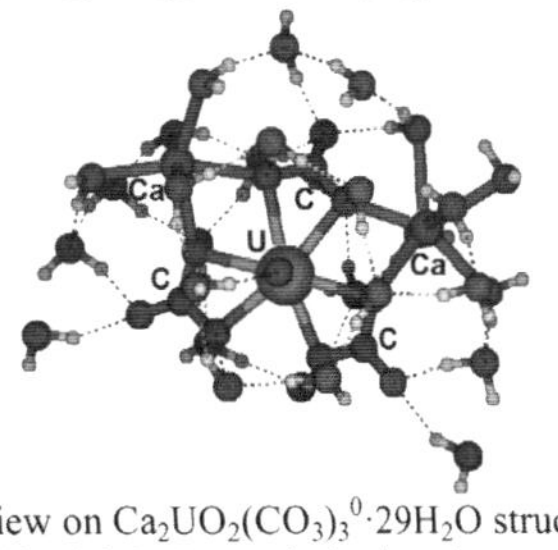

Fig. 33. Top view on $Ca_2UO_2(CO_3)_3^0\cdot 29H_2O$ structure (optimized within def-SVP atomic basis set)

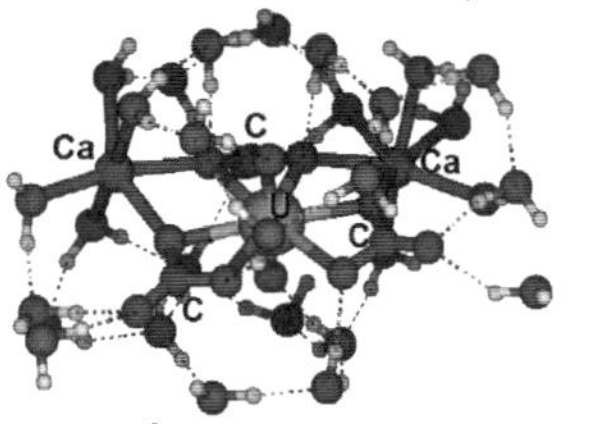

Fig. 34. $Ca_2UO_2(CO_3)_3^0\cdot 29H_2O$ structure (optimized within def-SVP atomic basis set)

Table 18: Calculated properties of $Ca_2UO_2(CO_3)_3^0.nH_2O$ models.

	b.	n	T_e	$\omega_{gs,1}$	$\omega_{gs,2}$	R
2	Tc	22		840		179
3	S	29	21311	832	842	180

7 Conclusions

Since s/n ratio in measured natural water samples has been very low (yet very low differences between all spectroscopic parameters of individual components (in few to tens cm^{-1}) imply the decomposition into individual components is ill-conditioned problem and s/n ratio should be rather high), for most samples, PARAFAC as a robust (exploiting correlations across wider set of data) chemometric method should be used. For eight component model (one component corresponds to artifact due to software background correction) individual component concentration profiles has been correlated to geochemical modelling prediction (by PhreeqC program). Assignment of individual component spectra and luminescence decay curves is presented in Tab. 8. In following Tab. 9 literature data on spectroscopic parameters of assigned chemical components are listed. From comparison of identical parameters in aforementioned tables we can conclude that internal consistency of PARAFAC data analysis is relatively high (in particular for luminescence life-times τ and symmetric mode stretching frequency of uranyl group ω_{gs}).

Samples S4 and S5 are specific since has been made by acidification of the sample S3 by HCl. In particular, sample S4 should have qualitatively different speciation than other samples (Fig. 19, for S4 species UO_2CO_3 and UO_2SO_4 should be dominant over ternary tris(carbonato) complexes $MnUO_2(CO_3)_3^{2n-4}$ (M=Ca,Mg, n=0,1,2)). This might be overlooked by PARAFAC, but should be visible in detailed interpretation of FATS results for the samples S3, S4 and S5.

And FATS method truly indicates one or two chemical components in S4 qualitatively different from the three seen as main constituents of all other samples – $Ca_2UO_2(CO_3)_3^0$, $CaUO_2(CO_3)_3^{2-}$ and $MgUO_2(CO_3)_3^{2-}$.

Quantum chemical modelling led to realistic values of vibrational frequencies and geometry of $Ca_2UO_2(CO_3)_3^0$ well comparable with literature, but excited state properties (including T_e) haven't been predicted with certainty useful in practical individual component assignment. The reason might be that most uranyl compounds has highly multi-reference character of excited states (excited states are not dominated by a single electron configuration as the UO_2^{2+} ground state) and more advanced, ab initio, method should be used (rather than TD-DFT). Spin-orbit coupling should be also included for excitate state properties (going beyond quasirelativistic approximation).

Eventhough, I hope to use rigorous chemometric methodic, the data are not enough to provide more certain insight into uranium speciation of given samples. But I hope that methodics presented as a road to qualitative results obtained is itself of great importance for future use.

I would support further experimental work on natural water samples with practical suggestions given below:

Acknowledgements

I would like to thank the Department of Nuclear Chemistry for support. the Czech Student Grant Agency (grant no. SGS16/250/OHK4/3T/14) for material support. Special thanks are for experimental collaborators from HZDR (Helmholtz-Zentrum Dresden-Rossendorf). Useful help and hints came from Dr. Šebesta, Fibich and Sláma. I would like to thank TESNAT 2017 Organizing Committee for hospitality and the opportunity to present my work. Access to computing and storage facilities: National Grid Infrastructure MetaCentrum (www.metacentrum.cz). under program CESNET LM2015042 is greatly appreciated.

References

1. R.J. Lakowicz, *Principles of Fluorescence Spectroscopy*, (Third Edition, Springer), (2006).
2. G. Meinrath, *Aquatic Chemistry of Uranium*, Review Focusing on Aspects of Environmental Chemistry, (Freiberg On-line Geoscience Vol.1), (1998).
3. http://www.world-nuclear.org/
4. W.G. Wright, J. Environ. Qual., **28**, No 4 (1999).
5. D. Vopálka, et al., J. Radioanal. Nucl. Chem., **286**: 681 (2010).
6. A. Vetešník, et al., *Uranium(VI) sulfate complexation as a function of temperature and ionic strength studied by TRLFS* In: Uranium, Mining and Hydrogeolology. Berlin: Springer-Verlag, p. 623-630. (2008).
7. J. Višňák, et al., *Study of the speciation in the system UO2(2+)-SO4(2-)-H2O by means of the UV-VIS spectrophotometry*, In The New Uranium Mining Boom edited by B. Merkel and M. Schipek, 643-652 (2012).
8. T. Vercouter, et al., Inorg. Chem., **47**, 2180-2189 (2008).
9. V. Sladkov, B. Fourest, F. Mercier, Dalton Trans., 7734-7740 (2009).
10. P. Lubal, J. Havel, Chem. Papers **51**, 213-220 (1997).
11. P. Lubal, J. Havel, Talanta **44**, 457-466 (1997).
12. M. Maeder, Y.-M. Neuhold, *Practical Data Analysis in Chemistry*, In Data Handling In Science and Technology, Advisory Editors: S. Rutan and B. Walczak, **26**, Elsevier, (2007).
13. J. Višňák, Master Thesis, FNSPE, Czech Technical University (2010) (In Czech).
14. A. de Juan, R. Tauler, Anal. Chim. Acta, **500**: 195–210 (2003).
15. R. Bro, et al., Applied Spectroscopy Reviews. 32 (3): 237–261 (1997).
16. Ch.F. Beckmann, S.M. Smith, Tensorial Extensions of Independent Component Analysis for Multi-Subject FMRI Analysis, FMRIB, Technical Report TR04CB1.
17. R. Harshman, M. Lundy, The PARAFAC model for three-way factor analysis and multidimensional scaling, In Research methods for multimode data analysis, Chap. 5, P. 122-215. Praeger, New York (1984).
18. R. Harshman, UCLA Working Papers in Phonetics, 16, 84 pp. (University Microfilms, Ann Arbor, No. 10,085) (1970).
19. G.H. Golub, C. Reinsch, Numerische Mathematik. **14**: 403–420, (1970).
20. P. Atkins, J. De Paula, *Physical Chemistry* (8th ed.). W.H. Freeman and Company. p. 212 (2006).
21. K.G. Dyall, K.Jr. Faegri, *Introduction to Relativistic Quantum Chemistry*, Oxford University Press, (2007).
22. P. Hohenberg, W. Kohn, Phys. Rev. **136**: B864–B871, (1964).
23. W. Kohn, L.J. Sham, Phys. Rev. **140**: A1133-A1138 (1965).
24. A. Szabo, N.S. Ostlund, Modern Quantum Chemistry: Introduction to Advanced Electronic Structure Theory, McGraw-Hill Publishing Company, New York, (1989).
25. E. Runge, E.K.U. Gross, Phys. Rev. Lett. **52**: 997–1000 (1984).

26. M. Petersilka, U.J. Gossmann; E.K.U. Gross, Phys. Rev. Lett. **76**: 1212–1215 (1996).
27. C. Ullrich, *Time-Dependent Density-Functional Theory: Concepts and Applications* (Oxford Graduate Texts), Oxford University Press, (2012).
28. A.D. Becke, J. Chem. Phys. **98**: 5648–5652 (1993).
29. F. Weigend et al, Chem. Phys. Letters **294**, 143 (1998).
30. TURBOMOLE V6.5 2013, a development of University of Karlsruhe and Forschungszentrum Karlsruhe GmbH, 1989-2007, TURBOMOLE GmbH, since 2007; available from http://www.turbomole.com.
31. F. Furche, et al., WIREs Comput Mol Sci, 4:91–100 (2014).
32. X. Cao, M. Dolg, J. Molec. Struct., 673, 203-209 (2004).
33. M. Dolg, in Proceedings, Second Edition, J. Grotendorst (Ed.), John von Neumann Institute for Computing, Julich, NIC Series, **3**, 507-540 (2000).
34. V. E. Jackson, et al., J. Phys. Chem. A, 112, 4095-4099, (2008).
35. Z. Wang, et al., Environ. Sci. Technol., **38**, 5591-5597 (2004).
36. J. Višňák, et al., *UO2(2+) - XO4(2-) - H2O TRLFS and spectrophotometric speciation study (X=S, Se)* In Lectures of colloquium on Radioanalytical methods, **IAA 14**, Edited by Jiří Mizera, Ioannes Marcus Marci Spectroscopic Society - IMMSS, ISBN 978-80-905704-4-3 (2014).
37. J.T. Bell, R.E. Biggers, *J. Molecular Spectroscopy*, **25**, 312 (1986).
38. L.H. Jones, Spectrochim. Acta, **10**, 395-403, (1958).
39. J.J. Olivero, R.L. Longbothum, J. of Quantitative Spect. and Radiative Transfer. **17**: 233–236 (1977).
40. G.K. Wertheim, et al., Review of Scientific Instruments. 45: 1369–1371 (1974).
41. M. Gal, et al., Spectrochim. Acta **48A** (1992).
42. L.C. Andrews, *Special Functions of Mathematics for Engineers*, Sec.Ed., SPIE Press, Oxford Science Publications, p. 110 (1998).
43. W.H. Greene, *Econometric Analysis*, Prentice-Hall, p. 926, fn. 11, (1993).
44. M. Born, R. Oppenheimer, Ann. Phys. **84**, 457 (1927).
45. J. Franck, Transactions of the Faraday Society, **21**: 536–542, (1926).
46. E. Condon, Phys Rev, **28**: 1182–1201, (1926).
47. A.S. Coolidge, et al., J. Chem. Phys, **4**: 193–211, (1936).
48. P.W. Atkins, R.S. Friedman, *Molecular Quantum Mechanics*, Oxford: Oxford University Press, (1999).
49. V.I. Minkin, Pure and Applied Chemistry. **71,** 10, 1919–1981.[g]
50. E. Rudberg, Diploma work in Physical Chemistry, Royal Institute of Technology, Sweden (KTH), (2003).
51. M. Abramowitz, I. A. Stegun, eds. "Chapter 22". *Handbook of Mathematical Functions with Formulas, Graphs, and Mathematical Tables*. Applied Mathematics Series. 55. Washington D.C., USA; Dover Publications. p. 773 (1983).
52. https://en.wikipedia.org/wiki/Hermite_polynomials
53. J. Višňák routine in Wolfram Mathematica [54].
54. Wolfram Research, Inc., Mathematica, Version 10.4, Champaign, IL (2016).
55. J.D.J. Ingle, S. R. Crouch, *Spectrochemical Analysis*, New Jersey: Prentice Hall, (1988).
56. J.-L. Chang, J. Mol. Spectrosc., **232**, 102-104, (2005).
57. R. Islampour, M. Dehestani, S. H. Lin, J. Mol. Spectrosc. **194**, 179–184 (1999).
58. P.-A. Malmqvist, N. Forsberg, Chem. Phys., **228**, 227–240, (1998).
59. K. Pierloot K., E. van Besien, J Chem Phys **123**, 204309 (2005).
60. J. Mehra, H. Rechenberg, *The Historical Development of Quantum Theory*, Volume 1 Part 2, (Springer, 2001).
61. G. Budylin, et al., *Optics Express*, **21**, No. 18, (2013).
62. H. D. Burrows, S. J. Formosinho, *J. Chem. Educ.*, **55** (2), p 125 (1978).
63. Laser-Induced Spectroscopy, Helmholtz Zentrum Dresden Rossendorf website, http://www.hzdr.de.
64. H.T. Evans Jr., *Science*, **141**, 154-158, (1963).

[g] The formula on the respective page 1941 there contains, however error/typo, the correct one should read

$$V = \frac{1}{2}\sum_{i,j=1}^{3N} \frac{\partial^2 V}{\partial q_i \, \partial q_j} q_i q_j . \qquad (66)$$

65. R.G. Denning, *Properties of the UO_2^{n+} (n=1,2) Ions*, Gmelin Handbuch der Anorganischen Chemie, Uran (Supplement) **A6**, 31-79, (1983).
66. R.G. Denning, et al., *Mol. Phys.*, **37**, 1089-1107 (1979).
67. B. Zhu, et al., *Microchem. J.*, **115**, 51-57, (2014).
68. C. Hennig, et al., *AIP Conf. Proc.*, **882**, 262 (2007).
69. J. Šebera et al, *In Scientific report of the project for SÚRAO, the Czech Radioactive Waste Repository Authority for the final control date (Phase 4)*, Department of Nuclear Chemistry, FNSPE, Czech Technical University, p 27-42 (In Czech), (July 2009).
70. J.N. Brønsted, *J. Am. Chem. Soc.*, **44**, 877–898 and 938–948 (1922).
71. G. Scatchard, *Chem. Rev.*, **19**, 309–327 (1936).
72. E.A. Guggenheim, *Applications of Statistical Mechanics*, Oxford: Clarendon Press, (1966).
73. M.R. Wright, *An Introduction to Aqueous Electrolyte Solutions*, section 10.6.15, Wiley, (2007).
74. I. Grenthe, H. Wanner, E. Osthols, *TDB-2, Guidelines for the extrapolation to zero ionic strength*, OECD Nuclear Energy Agency, NEA, Le Seine-St. Germain, (2000).
75. M. Bühl, et al., *Inorg. Chem.*, **50**, 299–308 (2011).
76. M. Zhou, et al., J. *Phys. Chem. A*, **104**, 5495-5502 (2000).
77. W.A. Gezahegne, et al., *Environ. Sci. Technol.*, **46**, 2228–2233 (2012).
78. B. Li, et al., *Comp. Theor. Chem.*, **1051**, 151–160, (2015).
79. L. Gagliardi, et al., *J. Phys. Chem. A*, **105**, 10602-10606, (2001).
80. P.D. Dau, et al., *J. Chem. Phys.*, **137**, 064315 (2012).
81. M. Head-Gordon, et al., *Chem. Phys. Lett.* **219**, 21 (1994).
82. P.A.M. Dirac, *Proc. Royal Soc.* (London) A, **123**, 714, (1929).
83. J.C. Slater, *Phys. Rev.*, **81**, 385, (1951).
84. S.H. Vosko, et al., *Can. J. Phys.*, **58** 1200, (1980).
85. C. Lee, W. Yang, R.G. Parr, *Phys. Rev. B*, **37**, 785 (1988).
86. A.D. Becke, *Phys. Rev.* A **38**, 3098 (1988).
87. J.P. Perdew, *Phys. Rev.* B **33**, 8822 (1986).
88. DIRAC, a relativistic ab initio electronic structure program, Release DIRAC13 (2013), written by L. Visscher, H.J. Aa. Jensen, R. Bast, and T. Saue.
89. A. Schäfer, et al., *Phys. Chem. Chem. Phys.*, **2**, 2187-2193, (2000).
90. K. Eichkorn, et al., *Chem. Phys. Letters* **242**, 652 (1995).
91. K. Eichkorn, et al., *Theor. Chem. Acc.* **97**, 119 (1997).
92. M. Sierka, et al., *J. Chem. Phys.* **118**, 9136, (2003).
93. S. Grimme, et al., *J. Chem. Phys*, **132**, 154104 (2010).
94. E.B. Wilson, Jr., *J. Chem. Phys.* **9** (1): 76–84 (1941).
95. A. Malijevský, *Statistic thermodynamics lessons*, University of Chemistry and Technology, Prague, 3.edition, 2009. http://old.vscht.cz/fch/cz/pomucky/ lekce.pdf (In Czech).
96. J. Višňák, A. Vetešník, K. Štamberg, J. Bok: Study of the Speciation in the System UO2(2+) – SO4(2-) – H2O by means of the UV-VIS spectrophotometry, *Annual report of the Department of Nuclear Chemistry 2011-2012*, Prague. http://www.jaderna-chemie.cz/data/documents/ vyrocni_zpravy/AR2011-2012.pdf.
97. J. Višňák, J. Bok, A. Vetešník, The Factor Analysis (Coupled with Speciation Computation) of Time Series as a tool for the Analysis of Spectra from the Time-Resolved Laser Fluorescence Spectroscopy, *Annual report of the Department of Nuclear Chemistry 2008-2010*, Prague..
98. J. Višňák, Laserem indukovaná fluorescenční spektroskopie s časovým rozlišením jako nástroj pro přímou speciaci uranu o nízkých koncentracích, Bachelor project, FNSPE, Czech Technical University in Prague (2008).
99. V. Sladkov, *J. Chromatography A*, **1276**, 120-125, (2013).
100. A. Schäfer, *J. Chem. Phys.*, **97**, 2571 (1992).
101. A.S.P. Gomes, et al., *Theor Chem Acc*, **127**: 369, (2010).
102. J.M. Martinez, L. Martinez, *J. Comput. Chem.*, **24**:819-825,(2003).
103. A.K. Rappé, et al., *J. Am. Chem. Soc.*, **114**(25), 10024–10035, (1992).
104. G. Schaftenaar and J.H. Noordik, *J. Comput.-Aided Mol. Design*, **14**, 123-134, (2000).
105. MATLAB and Statistics Toolbox Release 2012b, The MathWorks, Inc., Natick, Massachusetts, United States.
106. B. Drobot, et al., Chem. Sci., **6**, 964-972, (2015).

Quantum algorithms for computational nuclear physics revisited, particular case of second quantized formulation

Jakub Višňák[1,2,3*], *Petr* Veselý[4]

[1]Dep. of Nuclear Chemistry, FNSPE, Czech Technical University, Břehová 7, 115 19 Prague 1, Czech Rep.
[2]J. Heyrovský Institute of Physical Chemistry, Dolejškova 2155/3, 182 23 Prague 8, Czech Rep.
[3]Dep. of Chem. Phys. and Optics, FMP, Charles University in Prague, 121 16 Prague 2, Czech Rep.
[4]Nuclear Physics Institute, Czech Academy of Sciences, 250 68 Řež, Czech Rep.

Abstract. No core Full Configurational Interaction (NCFCI) calculations of Nuclear Bonding energy are resource demanding, in particular, computational time scales exponentially with the nucleon number A. In contrast to that, usage of quantum computers would allow an *efficient* (in polynomial time) NCFCI calculation and speed-up for other beyond-Mean-Field (correlation energy including) methods. To initiate feasibility studies of given quantum algorithms, we present an introduction to preliminary classical-computer simulation for the case of spherical nuclei (and ^4He in particular) within NCFCI with realistic chiral NNLO_opt potential.

1 Motivation

Due to the exponential scaling of possible configuration number, accurate NCFCI calculations done on classical computers have been limited up to the p-shell nuclei [1,2,14]. Quantum computers [3] could change that, allowing heavier nuclei calculations, since they offer exponential speed-up – to demonstrate that, Abrams-Lloyd algorithm [4-8] adopted to Computational Nuclear Physics realm will be presented followed by first computational steps towards greater feasibility study for the ^4He nuclei.

2 Problem formulation

2.1 Configuration interaction

General nuclear structure hamiltonian in second quantization (considering No-pair approximation and nucleonic structure with at most k-body interactions) has a form

$$\hat{H} = \sum_{n=1}^{k}\sum_{\mathbf{i}_n}\sum_{\mathbf{j}_n} V^{(n)}(\mathbf{i}_n,\mathbf{j}_n)\hat{a}^+(\mathbf{i}_n)\hat{a}(\mathbf{j}_n), \quad (1)$$

where $\mathbf{j}_n$ stands for n-tuple of integer indices ($j_1, j_2, \ldots, j_n$), $\hat{a}(\mathbf{j}_n)$ for a string of annihilation operators (2) and $\hat{a}^+(\mathbf{i}_n)$ for a string of creation operators (3), $V^{(n)}(\mathbf{i}_n,\mathbf{j}_n)$ is a complex coefficient (4),

$$\hat{a}(\mathbf{j}_n) = \hat{a}(j_1)\hat{a}(j_2) \quad \hat{a}(j_n), \quad (2)$$

$$\hat{a}^+(\mathbf{i}_n) = \hat{a}^+(i_n)\hat{a}^+(i_{n-1}) \quad \hat{a}^+(i_1), \quad (3)$$

$$V^{(n)}(\mathbf{i}_n,\mathbf{j}_n) = \left\langle i_1,i_2, \quad ,i_n \middle| \hat{V}^{(n)} \middle| j_1,j_2, \quad ,j_n \right\rangle, \quad (4)$$

the creation/annihilation refers to w complete orthonormal (ON) sets of one-particle functions (defined on spatial and spin-1/2 domains) – for an ordinary nuclei w=2, one set corresponding to protons, one to neutrons (e.g. we can reserve negative integers for the former and positive for latter). In the standard formulation, number of creation and annihilation operators for the given type of nucleon must be same in each term in (1). Creation and annihilation operators, as connected to two distinct types of fermions has to obey anticommutation relations (5) and commute with operators for the other nucleonic type

$$\begin{aligned}&\left[\hat{a}^+(i),\hat{a}(j)\right]_{sign(i\cdot j)} \equiv \\ &\equiv \hat{a}^+(i)\hat{a}(j)+sign(i\cdot j)\hat{a}(j)\hat{a}^+(i) = \delta(i,j)\end{aligned}, \quad (5)$$

$$\left[\hat{a}(i),\hat{a}(j)\right]_{sign(i\cdot j)} = 0. \quad (6)$$

While particular example tested consists of at most 2-body interactions (k = 2, the simplest non-trivial), the quantum algorithm presented can be generalized for any k (and any w). In (4) the complex coefficient is therefore identified with matrix element of n-body force operator in one-particle ON basis set. Due to the indistinguishability of identical nucleons and hermitian character of hamiltonian, coefficients $V^{(n)}$ must obey relations

$$V^{(n)}(\mathbf{i}_n,\mathbf{j}_n) = V^{(n)}(P\mathbf{i}_n,P\mathbf{j}_n) = \left(V^{(n)}(\mathbf{j}_n,\mathbf{i}_n)\right)^*, \quad (7)$$

where P is an arbitrary permutation of identical nucleon indices. For presented case of k = 2 and w = 2 ON set of Hartree-Fock (HF, Mean-field, [9-11]) solutions for the ground state of double magic nuclei has been used, starting with ^4He. The HF equations has following form

* Corresponding author: jvisnak@gmail.com

$$\hat{F}^{(p)}\left|\psi_i^{(p)}\right\rangle = \varepsilon_i^{(p)}\left|\psi_i^{(p)}\right\rangle, \quad (8)$$

$$\hat{F}^{(n)}\left|\psi_i^{(n)}\right\rangle = \varepsilon_i^{(n)}\left|\psi_i^{(n)}\right\rangle, \quad (9)$$

where Fockians ("p" for protons, "n" for neutrons) in arbitrary ON basis sets $\left\{\left|\phi_\mu^{(r)}\right\rangle \mid \mu \in I^{(r)}\right\}$ ($r \in$ {p, n}) has the form

$$\begin{aligned} F_{\mu\nu}^{(r)} &= \left\langle \phi_\mu^{(r)}\right| h^{(r)} \left|\phi_\nu^{(r)}\right\rangle + \\ &+ \sum_{\sigma \in occ(r)} \left\langle \phi_\mu^{(r)}\psi_\sigma^{(r)}\right\| V^{(rr)} \left\|\phi_\nu^{(r)}\psi_\sigma^{(r)}\right\rangle + \\ &+ \sum_{\sigma \in occ(s)} \left\langle \phi_\mu^{(r)}\psi_\sigma^{(s)}\right| V^{(rs)} \left|\phi_\nu^{(r)}\psi_\sigma^{(s)}\right\rangle, \end{aligned} \quad (10)$$

where $V^{(1)} = h^{(r)}$ represents operator of kinetic energy, second line on right side (10) is an antisymmetrized matrix element of proton-proton (for r=p) or neutron-neutron (r=n) interaction[a] operator and third line represent proton-neutron interaction energy operator matrix element (s=n if r=p and vice versa), the summations is over occupied ($0s_{1/2}$) one-particle states for given nucleon. Eq. (8) allows (1) to be expressed without $V^{(1)}=h^{(r)}$ matrix elements. There is also possibility to express (1) by (slightly transformed) $V^{(2)}$ elements only (see Supplementary Information). The connection between notations in (1-7) and (8-10) is

$$\left|\psi_i^{(r)}\right\rangle = \hat{a}^+(\rho \cdot i)\left|0\right\rangle, \quad (11)$$

where ρ = -1 for r = p and ρ = +1 for r = n, for the HF ground state wave-function

$$\left|HF\right\rangle = \hat{a}^+(k_Z)\hat{a}^+(q_{A-Z})\left|0\right\rangle, \quad (12)$$

where for ^{4}He, $k_Z = k_2$ = (-1,-2), $q_{A-Z} = q_2$ = (1,2) for the two lowest lying m-levels for each nucleon type.

For any realistic nucleon-nucleon potential there is no analytical solution for either HF eigen-problem or a general beyond-HF eigen-problem of the hamiltonian (1). The numerical solution is done by restriction to finite-dimensional one-particle subspaces (defined by index sets $I^{(p)}$, $I^{(n)}$). For most bounded state properties HF solution should serve as a mere ON one-particle basis set generating procedure and more realistic A-body wave-functions (respecting correlations between particle motions) should be obtained through diagonalization of (1) within (subspace) of product space of restricted one-particle state spaces for each particle. Due to the Ritz variational principle [12,13], eigenvalues obtained are upper bounds to eigenvalues of (1) with respect to non-restricted, complete space.

In case of (1) being diagonalized within all possible (appropriately anti-symmetrized) products of one-particle functions (Configurations) term Full Configuration Interaction (FCI) is used [11,14,15]. In case only certain configurations are used, the method is called Limited Configuration Interaction and possible variants could be CIS, CISD, CISDT, ..., for set of at most single, double, triple, ... excitations from ground state (or from set of model configurations in MRCI variant) [16]. FCI is size-consistent and therefore preferable over limited variants (even when calculated with respect to a greater one-particle sets), only FCI energy will converge with one-particle basis (limiting to completeness) to exact eigenvalue [16] of the original hamiltonian (1). The problem is, however, computational demands – since the number of configurations scales exponentially with nucleonic number, as $O(N^A)$, where N is size of one-particle basis set and A number of particles.

There are three possible classes of solutions to aforementioned scaling problem – 1. To account for excited configurations approximately (e.g. via an exponential Ansatz for Wave-operator in Couple Clusters methods [17-20]), parametrize multi-configurational wave-function in different way (e.g. as in Multi-configurational Self-Consistent Field MCSCF methods [16,21]), 2. To re-formulate bounded state problem within Density Functional Theory [22-24] or 3. Use a computational device itself based on rules of quantum mechanics [3-5,25-29] and therefore better suitable for eigen-value problem (1).

While the first two classes in fact avoid to calculate FCI energy itself, the only true solution is the last one. Proof-of-principle of a particular algorithm for a theoretical quantum computer device will be presented in the next section.

2.2 Abrams-Lloyd Algorithm and Iterative Phase Estimation

2.2.1 Introduction

While the first applications of Abrams-Lloyd algorithm [4] have been tailored for computational (*ab initio*) quantum chemistry [4-8] (e.g. FCI for electronic structure for fixed nuclei [26,38], FCI for electrons and nuclei in molecule treated on equal footing via the NOMO approach [30]), the possibility to use Abrams-Lloyd algorithm for computational physics problem has been also previously suggested (e.g. [31]).

Here the main steps of the algorithm will be briefly reminded. The algorithm aim is to estimate hamiltonian H eigenvalue E for given multiparticle stationary state based on representation of evolution operator $U = \exp(-i \Delta t H)$.

2.2.2 Inputs for algorithm

The inputs for algorithm consists of:

1. Bounds E_{min}, E_{max} to estimated eigenvalue, such that $E_{min} < E < E_{max}$ and desired accuracy in number of bits, m (the usual choices would be m = 17 to 27).
2. Estimate (initial guess) $\left|\Psi_0\right\rangle$ of corresponding eigenvector $\left|\Psi\right\rangle$ such that mutual overlap $S = \left|\left\langle \Psi_0 \middle| \Psi \right\rangle\right|^2$ is greater than S_{crit} (in an initial formulation $S_{crit} = \pi^2/16 \quad 0.62$)
3. Elements of second quantized hamiltonian (1)

[a] Please see Supplementary Information for the details. The two-body interaction contains also (second quantized form of) mass-polarization term $(1/M)\mathbf{p}_i \cdot \mathbf{p}_j$ ($\mathbf{p}_i$ being linear momentum of i-th particle) due to the CMS energy subtraction.

The quantum algorithm is probabilistic and provides any of the two closest binary approximates E_{bin} or $E_{bin}+2^{-m}$, such that the correct result E lies in between them ($E_{bin} < E < E_{bin}+2^{-m}$, or $E = E_{bin}+2^{-m}\,\delta$, where $\delta \in (0;1)$) with success probability [3,29] p_m

$$p_m \;\geq\; S \cdot h_m(\delta), \tag{13}$$ [b]

where overlap S has been defined in second point of input list and factor h depends on reminder δ and accuracy m (bits),

$$h_m(\delta) \;=\; \frac{1}{2^{2m}}\left(\frac{\sin^2(\pi\delta)}{\sin^2(\pi\delta/2^m)}+\frac{\sin^2(\pi(1-\delta))}{\sin^2(\pi(1-\delta)/2^m)}\right), \tag{14}$$

for any combination of δ and m, h has a lower bound,

$$h_m(\delta) \;\geq\; 8/\pi^2, \tag{15}$$

therefore, the condition $S > S_{crit}$ is important to assure that $p_m > 0.5$. By eventual repetition of quantum computer algorithm, the success probability can be then amplified as close to 1.00 as desired.

When the algorithm is just simulated for smaller number of one-particle basis states on a classical computer, the desired result and corresponding probability is well known and number of algorithm repetitions r to achieve probabilities of, e.g., 0.99, 0.9999 or 0.999999 can be computed[c] [30]. However, in case of real quantum computer, the basis size N (e.g. cardinality of one-particle basis sets) might be huge, not allowing classical computer calculation in realistic time, so success probability might not be known prior to calculation itself. Therefore, it is important to do parametric study suggesting that chosen initial guess $|\Psi_0^{(N)}\rangle$ will not drop under S_{crit} (e.g. based on extrapolation of overlap $S_N = |\langle\Psi_0^{(N)}|\Psi^{(N)}\rangle|^2$ from N values for which classical simulation is possible to N for which calculation will be done exclusively on quantum computer). Given extrapolated overlap S we can estimate p_m and number of repetitions r needed for result to be correctly decided from majority voting on (in each algorithm run) obtained binary expansions. In an unfortunate case, we would fail to choose correct eigenvector guess, the algorithm result is still not void, but provide us by the eigenvalue corresponding to eigenvector which has largest overlap with initial guess used.

[b] Even though this expression is a lower bound to probability, in most cases, the true probability isn't much higher than bound. The possible difference is due to the "overlap" of probability distribution peaks associated to different eigenvalues of studied hamiltonian H. In a typical case, the gap between eigenvalue in question and its neighbors is by orders larger than read-out accuracy (2^{-m}) and (13) can be considered as an equation.

[c] The upper bound to r could be decided through binomial distribution of one-bit-based outcomes "correct eigenvalue"/" incorrect eigenvalue". Then according to Chernoff bound [32-34], should be

$$r \;\leq\; d \cdot \frac{2\ln(10)\cdot p_m}{(p_m-1/2)^2}, \tag{16}$$

where overall success probability from majority vote on all r independent algorithm runs is $1\text{-}\varepsilon$, $\varepsilon = 10^{-d}$ (a typical choice for d = 2, 4 or 6 [30]). Wider spread of a register readouts allows for r smaller than upper bound on right side of (16) [30].

Possible initial guess should have to be representable *efficiently*, i.e. through at most polynomially (in one-particle basis set size) many unitary gates. The most straightforward choice would be Hartree-Fock (Mean Field) guess. However, for a multireference states (common for excited states but in case of small gap between highest occupied one-particle state and lowest unoccupied one or for open-shell systems possible for ground state as well) this choice might lead to $S < S_{crit}$ and more sophisticated initial guess would be recommended. First way would be beyond-Mean field eigenvector calculated with less computationally demanding method than FCI – e.g. CISD, CISDTQ, CCSD, MBPT or FCI [11,16] with excitations restricted to given, limited number of unoccupied (virtual) spin orbitals (Fockian eigenvectors) within same one-particle basis set generated by Hartree-Fock calculation. Second, different one-particle basis set (but connected with the one used in quantum algorithm via finite-dimensional unitary transformation) might be generated from MCSCF or as Natural Orbitals (NO) through diagonalization of one-particle density matrix from Many Body Perturbation Theory (MBPT) of some higher order [11,16]. Third, purely quantum approach would be the use of Adiabatic State Preparation (ASP) method [35], [4], which facilitates transition from ground state (represented in quantum register) of one hamiltonian (here Hartree-Fock hamiltonian) to ground state (represented in quantum register) of another hamiltonian (e.g. FCI hamiltonian (1)) under preposition that smooth transformation of hamiltonian in question do exist and that ground state and lowest excited state will neither exchange nor degenerate inside the process (the lowest gap between them then limits number of steps needed).

2.2.3 Algorithm description

Abrams-Lloyd algorithm is based on the Phase-Estimation algorithm [3,29] for calculation of eigenvalues and eigenvectors of unitary matrix U, represented as a quantum logical gate acting on l qubits of quantum register b for storage of the eigenvector $|\psi\rangle$ (Fig. 1), this register is initialized to initial guess $|\Psi_0\rangle$ as mentioned in previous section. Quantum register a is initialized to $|0\rangle|0\rangle \quad |0\rangle$ (m-times, for reference, please see the beginning of quantum wires on the left of Fig.1) state and serves to store bits $f_1, f_2, \ldots, f_m$ of computed eigenvalue (please see the measurement symbols on the end (rightmost part) of wires in Fig. 1), then E can be computed from linear equality

$$\frac{E-E_{min}}{E_{max}-E_{min}} \;=\; 0.f_1 f_2 \quad f_m + 2^{-m}\delta, \tag{17}$$

where $\delta \in (0;1)$, or $2^{-m}\delta$ is a reminder limiting accuracy. An alternative using only one qubit read-out register a exists and are discussed in [3,29-31,36-38] (Fig. 3 on page 3 [31]), as Iterative Phase Estimation A and B (IPEA A, IPEA B, [38]) – since m qubit needed is traded for m-times repetition of „decoupled" quantum circuit ([31], Fig. 3, page 3).

 DOI: 10.1051/epjconf/201715401030

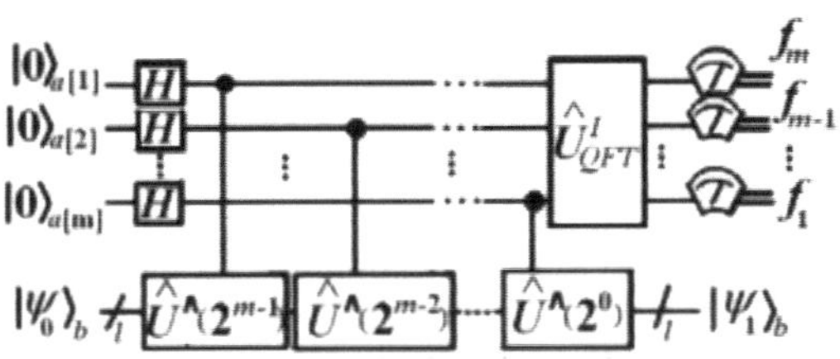

Fig. 1: Phase Estimation Algorithm (PEA) quantum logical circuit. Figure adopted from [31]

Inside the algorithm, the Hadamard gate,

$$H = \frac{1}{\sqrt{2}}\begin{pmatrix} 1 & 1 \\ 1 & -1 \end{pmatrix}, \tag{18}$$

is applied on each of m qubits of a register (boxes with H symbol in Fig. 1) creating equal weights superposition of binary numbers in a.

Then sequence of conditioned applications of quantum gate representation of U on register b is done (the lowest written wire in Fig. 1), with control qubit being from a register (note the corresponding full black circles in the middle of Fig. 1). This creates entangled state ((2) in [31]) where all x-th powers of U are entangled to corresponding binary representations of x in a register (19, first line).

The pre-last step of PEA is the inverse Quantum Fourier Transform (QFT, represented by a box with $\hat{U}^{I}_{QFT}$ in Fig. 1, the best-known versions has computational time cost of O(m log m) [39]) which in fact transform state vector in concatenated a+b quantum register (denoted |core> as in [31]) from time domain (time is discretely sampled in the a register) to the energy domain,

$$\begin{aligned} |\mathrm{core}\rangle &= \sum_t \exp(-i\,E_j\,t)c_j|t\rangle_a|\Psi_{(j)}\rangle_b \\ &\xrightarrow{QFT^I} \sum_E c_j|E\rangle_a|\Psi_{(j)}\rangle_b \end{aligned}, \tag{19}$$

where parametrization of unitary matrix U in question as $U = \exp(-\,\mathrm{i}\,\Delta t\,(H - E_{min}))$ (with $\Delta t = 2\pi/(E_{max} - E_{min})$ and H being FCI hamiltonain matrix) is supposed from now, E_j is exact FCI energy (minus E_{min}) for (j+1)-th state and c_j is the coefficient of expansion of initial guess $|\Psi_0\rangle$ to exact FCI eigenvectors $|\Psi_{(j)}\rangle$, i.e.,

$$S = |\langle\Psi_0|\Psi_{(0)}\rangle|^2 = |c_0|^2. \tag{20}$$

The last step is a measurement of all m qubits in a register, denoted by " " in Fig. 1 - to obtain m bits f_1, f_2, ..., f_m binary representation of energy through (17). Since in previous step, the quantum core is in an entangled state, where, however, amplitude of component entangled to desired binary representation of a correct energy $E_{(0)}$ should dominate, this step will lead to correct result with probability p_m given by (13).

At the end of the algorithm, the quantum register a is in state corresponding to measurement outcome and eigenvector in question is encoded inside the quantum register b as $|\psi_1\rangle_b$ (with the probability p_m it is the eigenvector we searched for). As wave-function is not an observable, there is no simple way how to read it from quantum computer qubit-based memory. However, we could exploit it for some subsequent quantum algorithm – e.g. recover as much as possible information through Quantum Tomography [40-41].

All sources of quantum computer power – *superpositioning* (Hadamard gate application), *parallelism* (ease to create all 2^m components in (19) in m steps), *entanglement* (the entangled state in (19)) and *destructive interference* (the inverse QFT) are exploited in PEA and this would lead to an exponential speed up for FCI energy calculation when compared to classical computer-based procedure... under an important preposition that both initial eigenvector guess and evolution operator U application (in general it is a l-qubit gate and should be decomposed into elementary one and two-qubit gates from universal quantum logic gate set [3]) will be represented efficiently (i.e. in number of time-steps (elementary gates) polynomial in one-particle basis set cardinality N). The latter preposition and U gate decomposition will be discussed in following section 2.2.4 and the initial sate preparation in 2.2.5

2.2.4 Computational complexity

The U, exponential of shifted and scaled hamiltonian (1), can be rewritten in a simplified form as

$$\hat{U} = \exp\left(i\,\Delta t \sum_{X=1}^{\Lambda}\hat{h}_X\right) = \exp\left(i\frac{\Delta t}{q}\sum_{X=1}^{\Lambda}\hat{h}_X\right)^q, \tag{21}$$

where h_X are hermitian combinations of terms from expansion (1), e.g. for some $X = X_0$,

$$\hat{h}_{X_0} = \hat{a}^+(i)\hat{a}^+(j)\hat{a}(k)\hat{a}(l) + \hat{a}^+(l)\hat{a}^+(k)\hat{a}(j)\hat{a}(i) \tag{22}$$

Furthermore, in accordance to the Trotter formula [3,37,42,43] for second order

$$\begin{aligned} &\exp(i(A+B)\Delta t) - o(\Delta t^3) = \\ &= \exp(i\,A\,\Delta t)\exp(i\,B\,\Delta t)\exp\left(\frac{1}{2}\left[\hat{A},\hat{B}\right]\Delta t^2\right), \end{aligned} \tag{23}$$

the expansion should be partitioned into largest commuting hermitian blocks to minimize number of terms for the next step. In this step, Trotter-Suzuki formula [3,37,42,43] is applied, with general η-th order form

$$\begin{aligned} &\exp\left(i\frac{\Delta t}{q}\sum_X \hat{h}_X\right) = \\ &= \prod_{k=1}^{2\Lambda\eta}\exp\left(i\frac{\Delta t}{q}\theta_{k,p(k)}\hat{h}_{p(k)}\right) + o\left(\left(\frac{\Delta t}{q}\right)^\eta\right), \end{aligned} \tag{24}$$

where $\theta_{k,p(k)}$ are real coefficients fulfilling $\sum_{k:\,p(k)=p}\theta_{k,p(k)} = 1$, Λ is number of terms in operator sum in (21) and p is projection from set $\{1,2,...,2\Lambda\eta\}$ *on* set $\{1,2,...,\Lambda\}$. For $\theta_{k,p(k)}$ value determination, please see [44],[3]. As discussed in [37], the order η should be

DOI: 10.1051/epjconf/201715401030

optimized. Too low order will provide an inaccurate approximation to the evolution operator U, yet too high order will lead to greater accumulated error (in classical computer simulation arising from rounding, in quantum computation from gate application error and decoherence). The optimization can be formulated (for an ideal quantum computer) as a minimization of ηq product (since the total number of exponential terms used to represent unitary operator U by elementary quantum gates is equal to $\Lambda\eta q$ and Λ is a fixed constant, for minimization technical details, see formula (296) in [37] and Fig. 7 there).

Then, each creation and annihilation operator in (22) or in any h_X term is represented via quantum gates acting on register a storing the eigenvector of U.

In case of the direct mapping the b quantum register qubits simply store the occupation numbers of m-levels (one-particle state characterized in Harmonic Oscillator scheme by fixed n, l, j and m quantum numbers) and since nucleons are fermions the binary character of qubits is well fit for this representation (occupation numbers are in {0;1}). Let us denote the total number of m-levels used in suggested calculation as N_m. As an example, wave-function representation for ^{4}He nuclei is

$$\left|\psi\left({}^4\mathrm{He}\right)\right\rangle = c_{HF}\underbrace{|1\rangle|1\rangle|0\rangle|0\rangle}_{\mathrm{N_m\ qubits\ for\ protons}}\ \underbrace{|1\rangle|1\rangle|0\rangle|0\rangle}_{\mathrm{N_m\ qubits\ for\ neutrons}} + \sum_i c_i \underbrace{\left|n_0^p\right\rangle\left|n_1^p\right\rangle \ \left|n_{N_m}^p\right\rangle}_{\mathrm{N_m\ qubits\ for\ protons}}\underbrace{\left|n_0^n\right\rangle\left|n_1^n\right\rangle \ \left|n_{N_m}^n\right\rangle}_{\mathrm{N_m\ qubits\ for\ neutrons}} ,\qquad (25)$$

where n_j^p is occupation number for j-th m-level for protons, n_j^n for neutrons, the sum runs over 2^(2N_m)-1 components, but only $\binom{N_m}{4}^2$ of them has non-zero c_i, or are further relevant (since we suppose no-sea approximation, $\Sigma_j n_j^p = \Sigma_j n_j^n = 2$). The size of b quantum registers $l = 2N_m$.

In the direct mapping case, creation and annihilation operators are represented by strings of Pauli matrices (Jordan-Wigner operator algebra mapping [45] employs O(N_m) gates, while more economic Bravyi-Kitaev mapping [46] needs just O(log N_m), please see section 2.5.2.1 in [31]) acting on individual qubits of b register.

While the direct mapping is qubit cost uneconomic (there are exponentially many more unphysical components in (25) than actively used in quantum computation!), the gate implementation is very straightforward and computational time is at most polynomial in N_m (if initial state preparation is *efficient*).

Computational time can be estimated by

$$t_Q \approx 2^m \cdot Te \cdot \frac{\Lambda}{N_p} \cdot N_1, \qquad (26)$$

where 2^m arise from PEA circuit (Fig. 1), $Te = \eta q$, stands for number of terms in Trotterization of $U = \exp(-\,\mathrm{i}\,\Delta t\, H)$, Λ stands for number of terms in second quantized hamiltonian (1), (21), N_p for the number of them to be executable in parallel and N_1 for the number of elementary gates needed to represent exponential of one single term h_X in (21). For general case of at most k-body forces in hamiltonian (1), $\Lambda = O\left(N_m^{2k}\right)$, since gates with non-overlapping target qubits could be applied in parallel, for Bravyi-Kitaev [46,31] mapping N_l=O(logN_m), N=O(N_m/logN_m) and for $\eta q = O\left(N_m^{\varepsilon}\right)$, where the power ε is shown to be at worst $\varepsilon = 6$ for two-body forces hamiltonian [47], but realistic estimates on molecular system provides $\varepsilon \in$ <1.5; 2.5> [48] the estimate (26) become

$$t_Q \approx O\left(\left(\log N_m\right)^2 \cdot N_m \wedge (2k-1+\varepsilon)\right), \qquad (27)$$

when compared to classical diagonalization time for FCI hamiltonian matrix which is

$$t_C \approx O(N_m \wedge (3A)), \qquad (28)^{\mathrm{d}}$$

considering the diagonalization has general scaling as O(n^3) for nxn matrix [49] (through for sparse matrices complexity decrease) and suggesting m-level scheme to be employed for classical computation as well (through that is ineffective) with upper bound to state number to be O(N_m^A) with nucleon number A. When slowly rising logarithmic term is omitted, powers in (27) and (28) can be compared and for

$$A \geq \frac{2k-1+\varepsilon}{3} \in (1.5;4), \qquad (29)$$

the quantum algorithm should be quicker than the classical one. However, since classical FCI diagonalization is done in j-scheme and FCI Hamiltonian sparsity is exploited, the denominator in (29) should decrease to 2 or 1.5, leading to critical A in interval (3; 8), where upper bounds are estimated with k = 3 and ε = 7. But it can be concluded that quantum FCI will be beneficial, if not from ^{4}He, from ^{16}O with certainty.

3 Numerical tests

In so far only classical FCI and limited CI preliminary calculations has been done for ^{4}He, In our calculations we use the nucleon-nucleon optimized chiral potential NNLO_opt. The coupling constants of this interaction were determined by a new optimization method in order to minimize the effects of the three-nucleon force [50], the results are summarized in the following table,

Table 1. Selected ^{4}He ground state (0^+) CI energies (E_{CI}) for NNLO_opt potential with Ω=16 MeV HO basis set parameter, Hartree-Fock (E_{HF}) and second order Many Body Perturbation Theory (E_{MBPT}), N denote maximal major shell number[e] and $|c_0|^2 = S$ the overlap between CI wave-function and Hartree-Fock ground state

N	E_{HF} [MeV]	E_{MBPT} [MeV]	CI type	E_{CI} [MeV]	$\lvert c_0\rvert^2$
1	-11,139290	-12,945510			

[d] Note that N_m might not be considered as independent to A. At least $N_m > A/2$ for a non-zero number of virtual one-particle states in FCI calculation. However, the N_m for fixed error in energy might not need to grow more than linearly in A, that is $N_m = \alpha A+\beta$ for some small positive α and β. Number of m-shells, N_m, is in relation to number of major shells, n, as $N_m \sim O(n^3)$, but number of j-shells as $N_j \sim O(n^2)$, therefore $N_j \sim O(N_m^{2/3})$, further explaining change of denominator in (29).

[e] $N = 2n+l$, where n starts from zero, e.g. N = 2 means collection of j-shells $0s_{1/2}$, $0p_{3/2}$, $0p_{1/2}$, $1s_{1/2}$, $0d_{5/2}$ and $0d_{3/2}$.

2	-13,542445	-16,280365	FCI	-17,590998	0,726
3	-13,542445	-19,754664	CISD	-20,116542	0,853
5	-15,496509	-24,637898			
8	-15,834727	-26,024630	FCI(2)	-16,188135	0,987
8			CISD(3)	-16,867806	0,971
	Exp. [51]	-28,296			

where round braces indicate major shell cut-off subsequent to Hartree-Fock basis generation. Hartree-Fock virtual one-particle states are in general rather problematic basis for correlation energy convergence as comparison of FCI for N=2 with FCI(2) for N=8 shows (or CISD for N = 3 compared to CISD(3) for N=8).

However, the overlap $S=|c_0|^2=|\langle HF|CI,0^+\rangle|^2$ and resulting PEA (quantum eigenvalue estimation) success probability estimates (Tab. 2) increase (e.g. from FCI to FCI(2)) as one-particle basis set (of a fixed cardinality) is generated through HF closer to HF limit.

Table 2. Phase Estimation Algorithm success probabilities for CI calculations from Tab. 1

N	CI type	p_{min}	p	p_{max}	$\lvert c_1\rvert^2/\lvert c_0\rvert^2$
2	FCI	0.588	0.702	0.726	$1.52\cdot10^{-3}$
3	CISD	0.691	0.828	0.853	$6.52\cdot10^{-3}$
8	FCI(2)	0.800	0.804	0.987	$1.02\cdot10^{-3}$
8	CISD(3)	0.787	0.845	0.971	$1.32\cdot10^{-3}$

As an example, $|c_1|^2/|c_0|^2$, relative configuration weight for the second most abundant configuration (for the first two rows in Tab. 2 corresponding to $0s_{1/2}\rightarrow 1s_{1/2}$ proton monoexcitation (for neutrons with by only 0.7% higher $|c_j|$), for the last two corresponding to biexcited configuration (one proton and one neutron from $0s_{1/2}$ do $0p_{3/2}$) is presented in Tab. 2 as well.

Bounds to probabilities, $p_{max}= S = |c_0|^2$, $p_{min} = S/(2S_{crit})$ and p are estimated from (13) with read-out parameter m=20 and δ derived from E_{max}=E_{HF} and $E_{min}\approx E_{HF}+\alpha(E_{MBPT}-E_{HF})$ (with α = 2). For a more detailed description of wave-function expansion into configurations for both ground and few chosen excited states, please see the Supplementary Information.

4 Conclusion

Preliminary classical CI calculations indicate quantum Abrams-Lloyd algorithm feasible for ^{4}He ground state FCI calculations, (at least for smaller basis set size) with the simplest single-determinantal initial eigenvector guess (identical to Hartree-Fock solution). For excited states, linear combination of a smaller set of configuration should be used as an initial eigenvector guess. As will be listed in following section, the feasibility studies should carry on for ^{4}He and later to be extended to larger nuclei. Different quantum algorithms than Abrams-Lloyd should be investigated as well.

5 Future prospects

1. To extend calculations to a greater basis set,
2. perform a full simulation classical of Abrams-Lloyd algorithm,
3. to provide reliable and efficiently computable lower bounds E_{min} (e.g. based on Tempel [52], Weinstein or Stevenson formulae [53] employing $\langle H^2\rangle$ or on Inner Projection technique [54]).

We would like to thank Dr. Libor Veis from J. Heyrovský Institute of Physical Chemistry in Prague and MSc. Michaela Chovancová from Department of Physics and Technology, University of Bergen for technical support and consultations, TESNAT 2017 Organizing Committee for hospitality and the opportunity to present this work. Access to computing and storage facilities: National Grid Infrastructure MetaCentrum (www.metacentrum.cz/en/index.html) under program CESNET LM2015042 is greatly appreciated.

References

1. B. Barrett, et al, Prog Part Nucl Phys **69** (2013) 131.
2. https://indico.cern.ch/event/609505/contributions/2517689/attachments/1448184/2231965/Mala_Skala_2017_Knapp.pdf
3. M.A. Nielsen, I.L. Chuang, *Quantum Computation and Quantum Information* (Cambridge University Press), (2000).
4. D.S. Abrams, S. Lloyd, Phys.Rev.Lett. **83**, 5162 (1999).
5. A. Aspuru-Guzik, et al, Science **309**, 1704 (2005).
6. H. Wang, et al, Phys. Chem. Chem. Phys. **10**, 5388-5393 (2008).
7. J.D. Whitfield, et al, Mol. Phys. **109**, 735 (2011).
8. L. Veis, J. Pittner, J. Chem. Phys. **133**, 194106 (2010).
9. P.-O. Löwdin, Phys. Rev. **97**, 1474 (1955).
10. D.J. Thouless, Nucl. Phys. **21**, 225 (1960).
11. J. Suhonen, *From Nucleons to Nucleus: Concepts of Microscopic Nuclear Theory*, Springer, Berlin, (2007).
12. W. Ritz, J. Reine, Angew. Math., **135** (1908) 1–61.
13. https://quantummechanics.ucsd.edu/ph130a
14. Q. Chong, J Phys Conf Ser **742** (2016) 012030.
15. Sternberg et al, Accelerating Full Configuration Interaction Calculations for Nuclear Structure in *SC '08: Proc. of the 2008 ACM/IEEE Conf. on Supercomputing*. https://www.osti.gov.
16. A. Szabo, N. S. Ostlund, Modern Quantum Chemistry: Introduction to Advanced Electronic Structure Theory, McGraw-Hill Publishing Company, New York, (1989).
17. G. Hagen et al, *Rep. Prog. Phys.* **77**, 096302 (2014).
18. O. Sinanoğlu (1962) J Chem Phys **36** (3) 706.
19. J. Čížek (1966) J Chem Phys **45** (11): 4256.
20. I. Shavitt, R.J. Bartlett (2009) *Many-Body Methods in Chemistry and Physics: MBPT and Coupled-Cluster Theory*. Cambridge University Press. ISBN 978-0-521-81832-2.
21. J. Lietz, etal., "Computational Nuclear Physics and Post Hartree-Fock Methods,", Lecture Notes in Physics (Springer), "*An advanced course in computational nuclear physics: Bridging the scales from quarks to neutron stars*", M. Hjorth-Jensen, M. P. Lombardo, U. van Kolck, Editors, (2016) [arXiv:1611.06765].
22. P. Hohenberg, W. Kohn, Phys. Rev. **136** (3B): B864–B871, (1964).
23. W. Kohn, L. J. Sham, Phys. Rev. **140** (4A): A1133-A1138 (1965).
24. J. Dobaczewski, J Phys Conf Ser **312**:092002(2011).
25. R. P. Feynmann, Int J Theor Phys, **21**, 6/7, (1982).
26. B. P. Lanyon et al, Nat Chem **2**, 106–111 (2010).
27. D. S. Abrams, S. Lloyd. Phys. Rev. Lett., **81**:3992–3995, (1998). quant-ph/9801041.
28. J. Gruska: *Quantum Computing*, Advanced Topics in Computer Science Series, The McGraw-Hill Companies, ISBN 007 709503 0.F (1999).
29. M. Dobšíček: *Quantum computing, phase estimation and applications*, Ph.D. thesis, Czech Technical University in Prague, Faculty of Electrical Engineering, Department of Computer Science and Engineering, (2008).
30. L. Veis et al Int J Quantum Chem (2016), **116**, 1328–1336. DOI: 10.1002/qua.25176.
31. J. Višňák, *EPJ Conferences* **100**, 01008 (2015).
32. H. Chernoff, (1952) Ann Math Stat **23** (4): 493–507.
33. T. Hagerup, C. Rüb, (1990) Inform Process Lett **33** (6): 305.
34. F. Nielsen, (2011). arXiv:1102.2684
35. L. Veis, J. Pittner, J Chem Phys, **140**, 214111 (2014).
36. A. Yu. Kitaev (1995). arXiv:quant-ph/9511026.

37. J. Višňák, *Quantum chemical algorithms for Quantum Computers*, Diploma Thesis, Mathematical-Physical Faculty, Charles University in Prague (In Czech, 2012).
38. L. Veis et al, Phys. Rev. A **85**, 030304(R) (2012).
39. L. Hales, S. Hallgren, Proc. of the 41st An. Sympos. on Foundations of Comp. Sci., p. 515, (2000),
40. R. Blume-Kohout. arxiv.quant-ph/0611080 (2006).
41. J. Řeháček, Z. Hradil, M. Ježek, M. (2001). Phys. Rev. A. **63**: 040303.
42. H. F. Trotter, Proc Am Math Soc **10**:545-551, (1959).
43. P. R. Chernoff, J Funct Anal, **2**:238-242, (1968).
44. Dobšíček M et al, Quantum Simulat ions: Report, Technical report, Microtechnology and Nanoscience, MC2, Chalmers, S-412 96 Göteborg, Sweden, (2010).
45. P. Jordan, E. Wigner, Z. Phys., **47**, Issue 9-10, pp 631-651 (1928).
46. S.B. Bravyi, A.Y. Kitaev, Ann. Phys. **298**, Iss. 1, 210-226(2002).
47. D. Wecker, B. Bauer, B.K. Clark, M.B. Hastings, M. Troyer, Phys. Rev. A **90**, 022305 (2014).
48. D. Poulin, et al, arXiv:1406.4920v1 (2014).
49. W.H. Press, S.A. Teukolsky, W.T. Vetterling, B.P. Flannery, (1992). Numerical Recipes in C (2nd ed.). Cambridge University Press. ISBN 0-521-43108-5.
50. A. Ekstroem et al (2013) Phys Rev Lett **110** 192502.
51. G. Audi *et al* (2017) Chinese Phys. C **41** 030001.
52. L. M. Delves (1972) *J. Phys. A: Gen. Phys.* **5** 1123.
53. Löwdin P.-O. and Quantum Chemistry Group (1965): Studies in Perturbation Theory. X. Lower Bounds to Energy Eigenvalues in Perturbation-Theory Ground State. Uppsala University, Sweden
54. F. Vinette, J. Čížek (1991) J Math Phys **32**, 3392.

AUTHOR INDEX